Laser
Spectroscopy

Proceedings of the XXII International Conference

Laser
Spectroscopy

XXII International Conference on Laser Spectroscopy (ICOLS2015)

Singapore, 28 June – 3 July 2015

Editor

Kai Dieckmann
NUS, Singapore

World Scientific

NEW JERSEY · LONDON · SINGAPORE · BEIJING · SHANGHAI · HONG KONG · TAIPEI · CHENNAI · TOKYO

Published by

World Scientific Publishing Co. Pte. Ltd.

5 Toh Tuck Link, Singapore 596224

USA office: 27 Warren Street, Suite 401-402, Hackensack, NJ 07601

UK office: 57 Shelton Street, Covent Garden, London WC2H 9HE

British Library Cataloguing-in-Publication Data
A catalogue record for this book is available from the British Library.

LASER SPECTROSCOPY
Proceedings of the XXII International Conference

ISBN 978-981-3200-60-9 (pbk)

Preface

The *22nd International Conference on Laser Spectroscopy (ICOLS)* took place in Singapore on June 28 - Jul 3, 2015 at the wonderful Shangri-La's Rasa Sentosa Resort, secluded from the centre of Singapore. The conference organization was supported by the Centre for Quantum Technologies (CQT), which is located at the National University of Singapore (NUS).

ICOLS features the latest developments in the area of laser spectroscopy and related topics in atomic, molecular, and optical physics and other disciplines. The talks covered a broad range of exciting physics, such as precision tests of fundamental symmetries with atoms and molecules, atomic clocks, quantum many-body physics with ultra-cold atoms, atom interferometry, quantum information science with photons and ions, quantum optics, and ultra-fast atomic and molecular dynamics.

The 2015 conference program was comprised of 14 sessions with 9 keynote addresses, 25 invited talks, and 3 hot topic talks. The speakers came from 15 different countries. A public evening talk by Alain Aspect at the Singapore Science Centre completed the program. The talks were attended by about 240 researchers including 66 students from 24 countries. I would like to thank the speakers and attendees, as well as the international program and steering committees and for contributing to a successful and inspiring conference. It is a pleasure to thank all members of the local organizing committee and the support team who made this conference possible. Their hard work always guaranteed a smooth organization and contributed to a very enjoyable atmosphere at the site.

Ever since the ICOLS conference series originated in 1973, its proceedings have been highly valued by many for capturing important developments in the field and offering the room to represent various aspects of specific research topics. As electronic media have been changing the possibilities for scientific publications the focus has been shifting away from the use of proceedings, as observed in recent issues of ICOLS. The present volume contains the texts of some of the invited talks delivered at the conference.

Kai Dieckmann   Singapore
(Conference/program chair)   June 14th, 2016
for the Series of International Conferences on
Laser Spectroscopy (ICOLS)

Organizing Committees

Local Organizing Committee
for the Series of International Conferences on
Laser Spectroscopy

Chair:
Kai Dieckmann   Centre for Quantum Technologies,
                       National University of Singapore

Co-Chair:
Murray Barrett   Centre for Quantum Technologies,
                       National University of Singapore

Local Organization:
Kwek Leong Chuan   Centre for Quantum Technologies,
                       Nanyang Technological University, Singapore

External Oversight:
Ken Baldwin   Laser Physics Center,
                       Australian National University, Canberra

Administration:
CQT admin team   The conference organization was supported by
                       the Centre for Quantum Technologies, Singapore

Conference secretariat:
Evon Tan   Centre for Quantum Technologies, Singapore

International Program Committee
for the Series of International Conferences on
Laser Spectroscopy

International Steering Committee
for the Series of International Conferences on
Laser Spectroscopy

Kenneth Baldwin (oversight) Australian National University, Canberra
Rainer Blatt.................University of Innsbruck, Austria
Wolfgang Ertmer............University of Hannover, Germany
Ted Hänsch .... Max-Planck-Institute for Quantum Optics, Garching, Germany
Peter Hannaford.............Swinburne University of Technology, Melbourne, Australia
Ed Hinds....................Imperial College, London, UK
Leo Hollberg .... Stanford University, USA
Siu Au Lee.................. Colorado State University, USA
Erling Riis .... University of Strathclyde, UK
Piet Schmidt .... PTB/University of Hannover, Germany
Vladan Vuletic .... Massachusetts Institute of Technology, Cambridge, USA
Mingsheng Zhan.............Wuhan Institute of Physics and Mathematics, Wuhan, China

Contents

1

Quantum Walks with Neutral Atoms:
Quantum Interference Effects of One and Two Particles

Carsten Robens, Stefan Brakhane, Dieter Meschede, and A. Alberti*

Institut für Angewandte Physik, Universität Bonn,
Wegelerstr. 8, D-53115 Bonn, Germany
* E-mail: alberti@iap.uni-bonn.de
http://quantum-technologies.iap.uni-bonn.de/

We report on the state of the art of quantum walk experiments with neutral atoms in state-dependent optical lattices. We demonstrate a novel state-dependent transport technique enabling the control of two spin-selective sublattices in a fully independent fashion. This transport technique allowed us to carry out a test of single-particle quantum interference based on the violation of the Leggett-Garg inequality and, more recently, to probe two-particle quantum interference effects with neutral atoms cooled into the motional ground state. These experiments lay the groundwork for the study of discrete-time quantum walks of strongly interacting, indistinguishable particles to demonstrate quantum cellular automata of neutral atoms.

Keywords: Quantum walks, Quantum transport, Optical lattices.

1. Introduction

The behavior of quantum particles in a periodic potential has been long investigated in physics. These studies allowed us to understand, for instance, the motion of electrons in crystal lattices. Since a few years, it has become possible to employ neutral atoms trapped in optical lattices to experimentally study the motion of

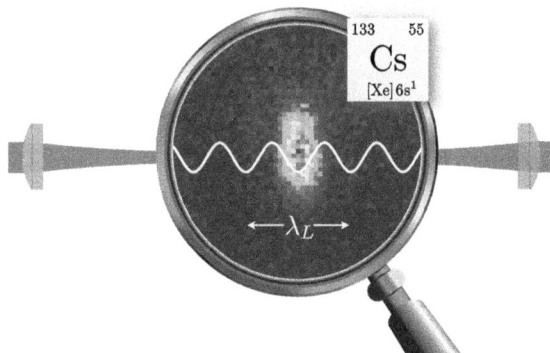

Fig. 1. Fluorescence image of a single atom in a one-dimensional optical lattice. The lattice constant is $\lambda_L/2 = 433\,\mathrm{nm}$. The atom image is not up to scale since the actual diffraction-limited size corresponds to four sites.

quantum particles in periodic potentials [1]. The common trait of these optical lattice experiments consists in tunneling through potential barriers, which allows matter wave to coherently delocalize in space [2]. A different approach to achieve coherent delocalization of matter waves is provided by discrete-time quantum walks (DTQWs). Instead of continuous tunneling through barriers, DTQWs rely on the controlled motion of a quantum particle, which is rigidly shifted in discrete steps conditioned on its internal degree of freedom, constituting a pseudo spin-1/2 system.

Quantum walks hold the promise to provide a universal computational primitive [3] and are the basic building blocks of a series of quantum algorithms [4]. Over the past few years, experimental implementations of quantum walks have been realized with cold atoms [5] and trapped ions [6, 7] with the pseudo spin-1/2 encoded in long-lived hyperfine states, as well as with photons spreading either through waveguide arrays [8] or fiber loop networks [9] with the pseudo spin-1/2 encoded in the polarization states, or even different spatial modes [10].

In our laboratory, we use single cesium atoms which are trapped in a very deep optical lattice potential, see Fig. 1. For the experimental realization, we need to coherently control the external degree of freedom (i.e., the atom's position in the lattice) as well as the internal one constituted by the atomic spin state [5]. Spatially resolved fluorescence detection allows us to measure the position with single site resolution [11] and to discriminate the two spin states $|\uparrow\rangle$ and $|\downarrow\rangle$ with good fidelity via the so-called push-out method [12]. More recently, we demonstrated that state-dependent optical lattices can be used to perform projective measurement of the atom's spin state even without relying on the push-out method. This technique plays a central role in the realization of interaction-free measurement to falsify classical trajectory theories (see Sec. 4).

2. State-dependent optical lattice

We use the outermost cesium hyperfine levels — namely $|\uparrow\rangle := |F = 4, m_F = 4\rangle$ and $|\downarrow\rangle := |F = 3, m_F = 3\rangle$ — to realize the pseudo spin-1/2 system. The two levels can

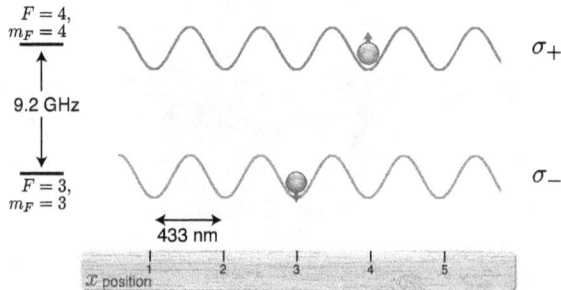

Fig. 2. State-dependent optical lattices acting selectively on either one of two long-lived hyperfine states of a cesium atom. Upper and lower lattices originate from σ_+ and σ_- circularly polarized standing wave light fields, respectively.

be coupled by microwave radiation at $9.2\,\text{GHz}$. Furthermore, due to the different ac-polarizability of these hyperfine levels, a magic wavelength exists at $\lambda_\text{L} = 866\,\text{nm}$ that enables spin-dependent optical potentials [13]: Atoms in spin $|\uparrow\rangle$ $(|\downarrow\rangle)$ state couple with light with σ_+ right-handed $(\sigma_-$ left-handed) circular polarization, yielding the following potentials:

$$U_\uparrow(t) = U_\uparrow^{(0)} \cos\{2k_\text{L}[x - x_\uparrow(t)]\} \quad \text{and} \quad U_\downarrow(t) = U_\downarrow^{(0)} \cos\{2k_\text{L}[x - x_\downarrow(t)]\} \quad (1)$$

where $U_{\uparrow,\downarrow}^{(0)}$ is the lattice depth that can be individually controlled in the experiment for both spin states, as well as the individual position $x_{\uparrow,\downarrow}(t)$ of the two periodic potentials.

Since we work in a deep optical lattice, such that tunneling between lattice sites is fully negligible, the trajectory of an atom in the $|\uparrow\rangle$ $(|\downarrow\rangle)$ state is determined by the motion of its σ_+ (σ_-) sublattice. While in earlier experiments (e.g., [5, 14]) the translation of each lattice potential was realized by an electro-optical retardation plate, we have recently developed a novel method, which synthesizes the state-dependent lattices from two independent σ_+ and σ_- optical standing waves, see Fig. 2. The relative position between the two standing waves is controlled by an optoelectronic servo loop with a resolution on the order of $\lambda_L/5000$ and a bandwidth of approximately $500\,\text{kHz}$. The bandwidth is primarily limited by the finite response time of acousto-optic modulators, which are employed to stabilize the relative position of two sublattices. The state-dependent optical conveyor belt allows us to transport atoms arbitrarily over tens of lattice sites, as we demonstrated in one of our experiments falsifying classical trajectory theories [15] (see Sect. 4).

3. Discrete-time quantum walks

Discrete-time quantum walks are the quantum analog of random walks. In the classical world, the walker decides at discrete time steps whether to move one site leftward or rightward depending upon the result of tossing a coin — heads or tails. As shown in Fig. 3(a), a quantum walker, instead, is put at every time step in a

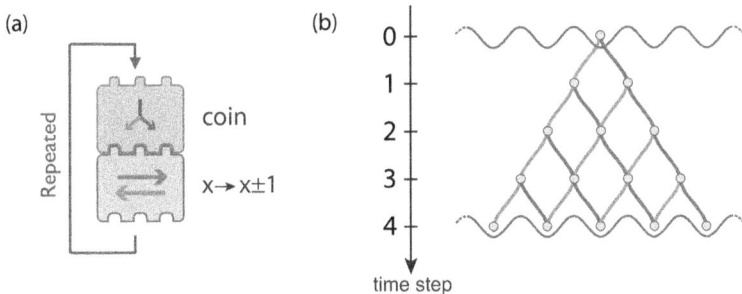

Fig. 3.   Discrete-time quantum walks in position space. (a) Discrete unitary operations defining the quantum walk's step. (b) Delocalization of the quantum walker over multiple paths. The number of paths scales increases exponentially with the number of time steps.

4

coherent superposition of two internal states (coin operation), and it is subsequently shifted by one lattice site in a direction subject to the spin state (spin-dependent shift operation), e.g., $|\uparrow\rangle$ to the left and $|\downarrow\rangle$ to the right.

The coin operation is experimentally realized by using microwave radiation that resonantly couples the two hyperfine states. This allows us to achieve any arbitrary unitary transformation of the pseudo spin-1/2 with the coin operation. The most frequently used coin, however, is the Hadamard coin, which produces an equal superposition of the two spin states (coin angle equal to $\pi/2$). The spin-dependent shift operation is realized by employing our conveyor belt transport technique, which moves the atom by one site rightward or leftward depending on the internal state. After applying both operations the trajectory of an atom will thus be split, giving rise to a beam splitter operation of a single atom interferometer [16].

After iteratively repeating the coin and shift operation, the matter wave spreads over multiple trajectories in position space, as illustrated in Fig. 3(b), producing a complex multi-path interference effect. The resulting probability distribution measured after a twenty-step quantum walk is shown in Fig. 3(a), where originally the walker was prepared in site 0 with spin $|\uparrow\rangle$ state. The prominent peak on the left-hand side provides signature of multi-path interference. Furthermore, the quantum walk spreads ballistically with the number of time steps n in contrast to a classic random walk, which spreads diffusively with a Gaussian distribution of width $\sqrt{n}$. Decoherence reduces the interference contrast, turning the quantum walk into a classical random walk. The width of the measured probability distribution is a useful analysis tool to discriminate quantum walks from classical random walks. Fig. 4(b) shows the measured RMS width for an increasing number of time steps, exhibiting ballistic spreading up to a few tens of steps. We investigated more than ten different physical decoherence mechanisms, which can be divided into two classes

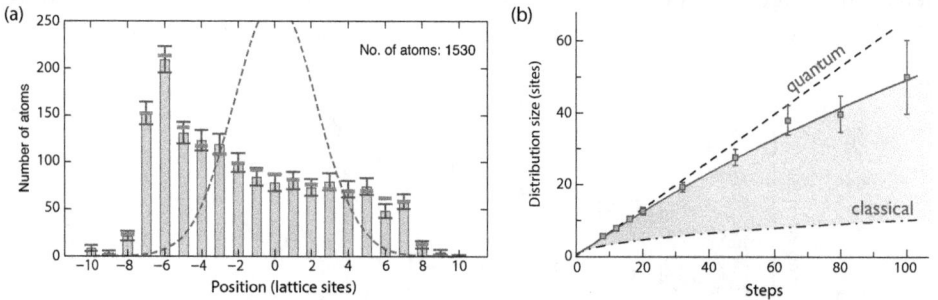

Fig. 4. Discrete-time quantum walks of single Cs atoms. (a) Probability distribution of single atoms after 20-step quantum walks. Bars with confidence intervals are the experimental data, short horizontal lines are the theoretical prediction with about 5% coherence loss per step. Dashed line is the prediction for a random walk (no coherence). (b) Size of the probability distribution as a function of the number of steps. Ballistic transport (quantum) and diffusive transport (classical) are shown as asymptotic cases.

depending on whether they couple with the spin or positional degree of freedom [17]. The short horizontal lines in the figure represent the decoherence model of quantum walks, indicating a good agreement with the experimental data. With regard to the experimental data in Fig. 4(b), the number of coherent steps is primarily limited by decoherence arising from light shifts, which however, is expected to vanish with the atoms cooled to the three-dimensional ground state of the optical lattice, see Sect. 5.

We exploited the possibility of delocalizing atoms over tens of lattice sites to study the physics of a charged particle, for example an electron, in a periodic potential under the effect of an external homogeneous force. Predicted by Felix Bloch nearly 90 years ago, the quantum particle in the lattice, instead of being indefinitely accelerated, performs periodic oscillations. We have shown experimentally that electric quantum walks do exhibit a similar behavior as Bloch oscillations, where both sublattices are accelerated at each time step for a short time interval to reproduce the action of an external electric field [14]. However, due to the time discreteness of the electric field operation (sublattices' acceleration), an even richer range of quantum transport regimes spanning from coherent delocalization to dynamical (Anderson-like) localization has been predicted [18] and observed.

4. Falsifying classical trajectory theories

The superposition principle is one of the pillars of quantum theory, which goes beyond the classical concept of particles moving along well defined trajectories. To understand the motion of a quantum particle, instead, quantum theory takes into account all possible trajectories that the particle can take. This idea lies at the heart of the quantum path integral formalism [19]. Up until today it is an unresolved question how to reconcile the quantum mechanical worldview, where physical objects obey the unitary Schrödinger equation, with the macro-realistic one, where objects are in one definite state at all times. For instance, the macroscopic apparatus of a Stern-Gerlach experiment always measures a definite orientation of the electron's spin, although the electron is, according to quantum mechanics, in a superposition of both spin orientations. To explain such a wave function reduction to a definite state, different ideas have been put forward, which can be coarsely divided into two groups [20]: (a) Interpretational solutions like, among others, the decoherence approach, Bohmian mechanics, and many-worlds theory. (b) Objective collapse theories such as continuous spontaneous localization, and gravitational collapse theory. What distinguishes theories of type (b) is the assumption of an objective reduction of superposition states involving mechanical degrees of freedom (i.e., massive particles), which deviates from a Schrödinger-type quantum dynamics [21].

In 1985 Leggett and Garg (LG) derived an inequality relating correlation measurements performed at different times, providing an objective criterion to distinguish between quantum (a) and macro-realistic (b) theories [22]. The inequality is

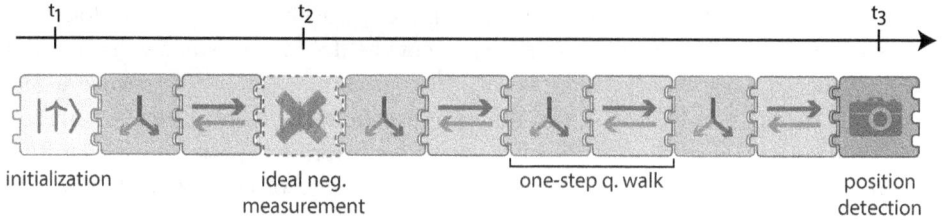

Fig. 5. Discrete-time quantum walk of four steps violating the Leggett-Garg inequality. Measurements are performed at three different times t_i. Only the position measurement at t_2 must be performed according to the ideal negative measurement protocol of Fig. 5. The measurement consists in spin selectively removing atoms in spin $|\downarrow\rangle$. Alternatively, atoms in spin $|\uparrow\rangle$ are selectively removed (not shown in the figure).

derived under two assumptions that embody the macro-realistic worldview, (A1) macro-realism (i.e., massive particles follow classical trajectories) and (A2) non-invasive measurability (i.e., the position of a macroscopic object can be measured without perturbing its subsequent evolution) [15]. Hence, the experimental violation of LG inequality falsifies a macro-realistic description of the studied phenomenon. Our experimental apparatus offers an ideal platform to put classical trajectories theories of type (b) to the test, as we are capable of observing the position of a massive particle, namely an atom, moving through controlled trajectories in a one-dimensional optical lattice.

The LG inequality binds the linear combinations of two-time correlation measurements according to

$$K = \langle Q(t_2)Q(t_1)\rangle + \langle Q(t_3)Q(t_2)\rangle - \langle Q(t_3)Q(t_1)\rangle \leq 1\,, \qquad (2)$$

where $Q(t_i)$ are measurements performed at three different times t_i which are bound by $|Q(t_i)| \leq 1$, but can otherwise be freely defined. We aim to disprove macro-realistic interpretation of our DTQWs by considering a four-step quantum walk as shown in Fig. 5, where we choose $Q(t_i)$ to be a function of the measured particle position x [15]:

$$Q(t_1) = +1\,, \qquad (3)$$

$$Q(t_2) = \begin{cases} +1 & \text{if} \quad \hat{x} = +1 \\ +1 & \text{if} \quad \hat{x} = -1 \end{cases}\,, \qquad (4)$$

$$Q(t_3) = \begin{cases} +1 & \text{if} \quad \hat{x} > 0 \\ -1 & \text{if} \quad \hat{x} \leq 0 \end{cases}\,, \qquad (5)$$

where x is position of the atom in units of lattice sites with respect to the initial position $x = 0$ at time t_1. Note that at time t_2 we assign the same value regardless of the measured position. This is, in fact, entirely consistent with the LG inequality and helps remark the importance that a measurement is at all performed at time t_2.

This measurement also represents the most challenging aspect of an experimental violation of LG inequality. To avoid invalidating hypothesis (A2) by experimental inadvertence, Leggett and Garg themselves introduced the idea of an ideal negative measurement protocol, which is consistent with the macro-realistic hypothesis (A1). The idea is based on a classical measurement protocol, which consists in performing an interaction-free measurement of position as illustrated in Fig. 6. While from a macro-realistic perspective, ideal negative measurements are non-invasive (hence, consistent with (A2)), it is apparent that from a quantum mechanical point of view the measurement $Q(t_2)$ causes a projection of the wave function to a definite trajectory, which in turn conditions the subsequent motion.

In the experiment, the ideal negative measurement at time t_2 is implemented by selectively relocating atoms in, e.g, spin $|\uparrow\rangle$ state so far away that they are effectively removed from the system. Since position and spin after the first step (time t_2) are perfectly correlated, a measurement of spin is equivalent to one of position. Finding the particle at time t_3 allows us to infer its position at time t_2 (in the example above: right site, $|\downarrow\rangle$). We also remark that the measurement at time t_3 does not need to be carried out with the ideal negative measurement protocol since the atom's evolution after t_3 is not relevant. Experimentally, we performed ideal negative measurements utilizing a novel spin-dependent conveyor belt technique, which allows us to shift only one sublattice at a time over multiple lattice sites, while leaving the other at rest.

The measured probability distributions shown in Fig. 7(a,b) show distinct profiles depending on whether a measurement has been performed at time t_2 or not. This difference gives rise to a violation of LG inequality when the measured probability distributions are used to compute the correlation function K in Eq. (2). Quantum mechanically, it is also understood that the different outcome is related to the measurement process at t_2, which in spite of being interaction-free, causes a projection of the wave function to

Fig. 6. Ideal negative measurement protocol. (a) A macro-realistic cat is prepared in an unknown state either under the left or the right container. (b) Detecting the absence of the feline in the right container enables us to measure its position without any direct interaction. (c) Measurements that directly detect its presence could have disturbed the animal by direct interaction. These measurements must thus be discarded by post-selection.

a statistical mixture. Fig. 7 shows the value of K as a function of the coin angle θ, which exhibits a maximal violation of 6σ for Hadamard walk ($\theta = \pi/2$) when the coin operation prepares an equal superposition of both spin states. Only two particular walks ($\theta = 0$ and $\theta = \pi$) fulfill the LG inequality, since in these cases the coin operation creates no superposition states.

Although our test mass — the cesium atom — is unquestionably microscopic, this experiment represents the most macroscopic test of quantum superposition states based on the stringent criteria provided by the LG inequality. Many regard the LG inequality as the gold standard to discern quantum superposition states, in like manner the Bell inequality has become the widely accepted criterion to test non-locality. Our experiment lays the groundwork for future tests with increasing "macroscopicity" [21] that could shed light on quantum to classical transition.

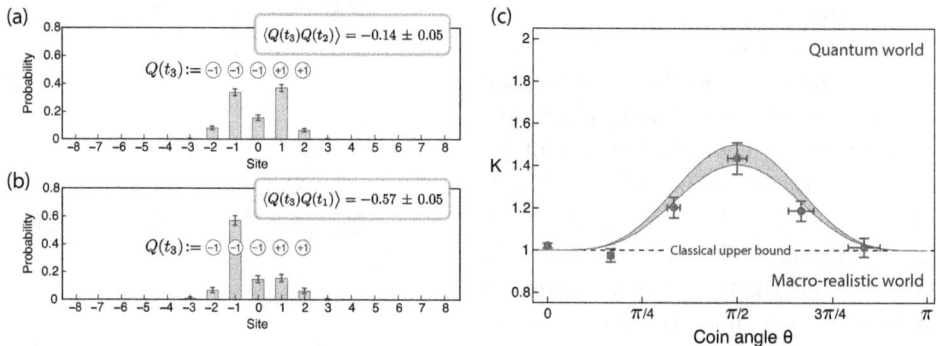

Fig. 7. Leggett-Garg test of quantum superposition principle. Measurement of the Leggett-Garg correlation functions with (a) and without (b) the ideal negative measurement at time t_2. (c) Maximal violation (by 6σ) of the Leggett-Garg inequality in Eq. (2) is measured for the Hadamard quantum walk (coin angle $\theta = \pi/2$).

5. Three-dimensional ground-state cooling

Over the years, our quantum walk experiments have demonstrated a remarkable control of single quantum particles in optical lattices. The next frontier consists in exploring the motion of strongly correlated quantum particles [23]. A necessary prerequisite for these experiments is that atoms must be cooled to the lowest three-dimensional (3D) vibrational state of a single lattice site, in order to confine their motion to a very small volume and let them collide in a controllable way.

After molasses cooling, our cesium atoms trapped in the optical lattice occupy different 3D vibrational quantum states, with a statistical occurrence given by the Boltzmann distribution. Yet, atoms can be subsequently prepared in the ground state by using resolved sideband cooling techniques to attain the 3D vibrational ground state [24]. Along the longitudinal direction of the optical lattice, we reliably employ microwave sideband cooling [25, 26], which allows us to achieve a ground-state population of around 99% starting from an initial population of around 50%. Sideband cooling, however, cannot be directly employed in the transverse direction because of the weak transversal confinement with trap frequencies on the order of a few kHz (compared to the longitudinal trap frequency of > 110 kHz), which are on the same order or even smaller than the recoil frequency of a scattered photon (≈ 2 kHz for Cs atoms). This corresponds to a Lamb Dicke parameter $\eta \gtrsim 1$. To circumvent this problem we superimposed an additional blue-detuned hollow laser beam (often referred to as "doughnut beam") with the one-dimensional optical lattice, which increases the transversal trap frequency up to 20 kHz corresponding to $\eta \approx 0.3$. Fig. 8 shows fluorescence images of single atoms, demonstrating the transverse compression produced by the superimposed doughnut beam.

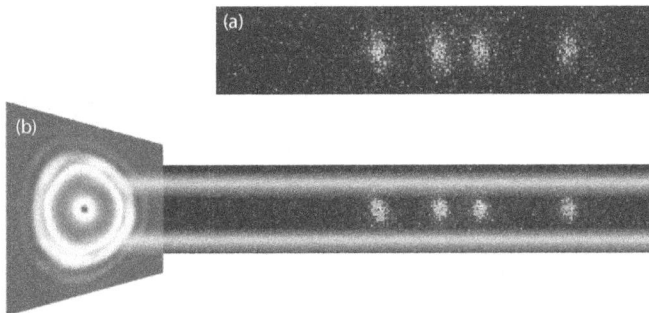

Fig. 8.   Transverse compression of atoms by hollow trap. (a) Without extra transverse confinement, thermal atoms are elongated in the transverse direction (y-axis in the figure). (b) By adding a doughnut-shaped blue-detuned laser beam, atoms' motion is transversally squeezed. The radius of the round shape is essentially limited by the 2 μm optical resolution of the microscope's objective lens (NA=0.23).

We use a single pair of Raman beams (one collinear with and the other perpendicular to the optical lattice) to cool both transverse directions according to the Raman cooling scheme in Fig. 9(a). In fact, we exploit a slight ellipticity (on the percentage level) of the doughnut trap to ensure motional coupling between the two transverse directions, so that the momentum transfer provided along a single direction by the Raman transition suffices to cool the atomic motion in both transverse directions. Fig. 9(b) shows an exemplary Raman sideband spectrum, where the suppression of the first blue sideband demonstrates transverse ground state cooling of atoms. By analyzing the relative heights of the sideband peaks, we infer a transversal ground-state population of 85% per direction, starting from an initial population of around < 1%. This leads to an overall 3D ground-state population of about 65%.

(a)

(b)

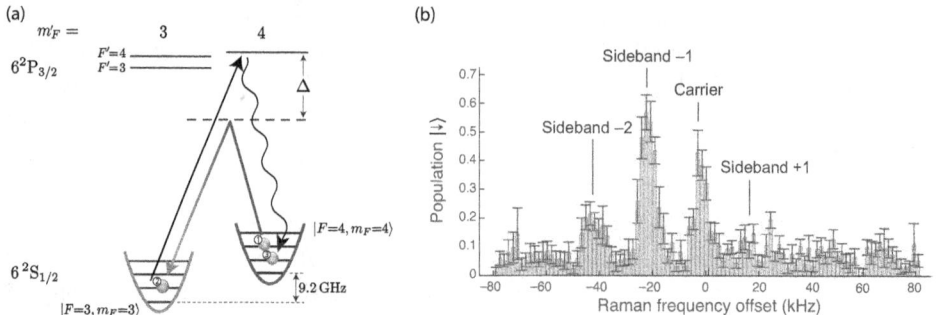

Fig. 9. Raman sideband cooling of single Cs atoms to the the 3D ground state. (a) Level scheme employed to cool Cs atoms. Circled numbers denotes the evolution of one atom during one cooling cycle. (b) Nearly-full suppression of the first blue transverse sideband indicates high occupancy of the ground state by the laser-sideband-cooled atoms. The blue longitudinal sideband at around 100 kHz is also highly suppressed (not shown in the figure).

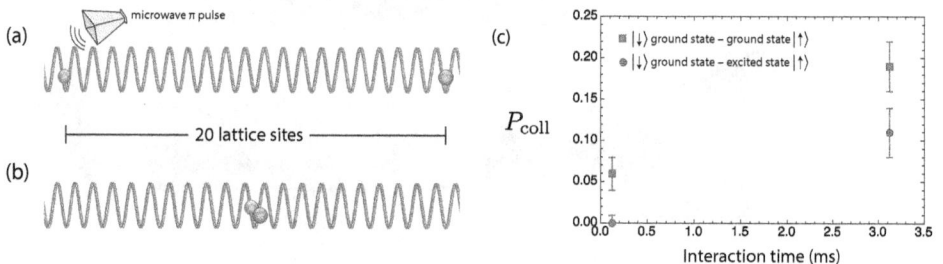

(a)

(b)

(c)

Fig. 10. Probing two-atom collisions in state-dependent optical lattices. (a) Two atoms are initially placed at a relative distance of 20 sites and cooled to the motional ground state. Using an addressing magnetic field gradient, one of the two atoms has its spin flipped by a microwave π pulse. (b) Both atoms are transported to the same lattice site by a single adiabatic spin-dependent shift operation lasting $\approx 500\,\mu$s. (c) Based on the model in equation (6), we determine the probability P_{coll} for two atoms to be lost due to a hyperfine changing collision. By exciting selectively the atom in $|\uparrow\rangle$, we observe a reduced probability of collisional losses.

6. Probing two-atom collisions at high densities

The capability to control the position of individual cesium atoms with high fidelity combined with 3D ground state cooling gives us the possibility to study the motion of strongly interacting particles. In addition, our system provides an ideal platform to measure atomic properties such as the scattering length between different spin combinations using exactly two atoms.

We carried out first experiments reporting on collisional losses due to inelastic collisions occurring at high two-atom densities. In previous experiments [27] it was shown that a dense cloud of cesium atoms undergoes hyperfine state-changing collisions on a time scale of a minute for densities $n \approx 10^{10}\,\text{cm}^{-3}$. Our experimental apparatus allows us to achieve densities six orders of magnitude higher by transporting two 3D-cooled atoms into the same lattice site according to the scheme illustrated in Fig. 10.

The energy released in the inelastic collision leads to losses of both atoms. By recording the occurrences at which both, one, or no atom remains in the optical lattice after a variable interaction time, we can extract the probability of inelastic collisional losses P_{coll} using a simple model:

$$\text{Probability that} \begin{cases} \text{no atom survives} = (1-P)^2 \, P_{coll} + P^2 \\ \text{1 atom survives} = 2P(1-P) \\ \text{both atoms survive} = (1-P)^2(1-P_{coll}) \end{cases} \quad (6)$$

where $1-P$ is probability for a single atom to remain trapped in the optical lattice during the experimental sequence in the absence of collisions. Independent measurements show that $1-P \approx 91\%$, which is mainly limited by technical reasons (timing of the experimental sequence) and additional losses experienced due to transverse cooling. Experiments with tighter transverse confinement are expected to reach single-atom survival probabilities close to 99%.

These preliminary results exhibit losses detectable already for interaction times on the ms scale, shown by the squares in Fig. 10(c). We can furthermore verify that the inelastic collision probability P_{coll} depends on the two-atom density. For that purpose, we excite the atom in $|\uparrow\rangle$ with a spin-dependent shaking of the σ_+ sublattice, which increases the volume of the atom's wave function. The reduced probability of collisional losses in this case is shown by the circle points in the same figure.

7. Microwave Hong-Ou-Mandel interferometer with massive particles

Ultracold atoms in the vibrational ground state of the lattice potential allow us to explore fascinating quantum-mechanical interference effects between two (or more) indistinguishable neutral atoms. Quantum mechanics shows that quantum correlated states of two particles can be produced even if particles are non interacting. The most prominent example is provided by the Hong-Ou-Mandel (HOM) experiment [28]. It demonstrated that two indistinguishable photons (with identical polarization and transverse mode) impinging simultaneously upon a beam splitter emerge in a quantum entangled state, where both photons exit from the beam splitter either through one or the other output port, but not from separate ones. The quantum correlation results from quantum interference of two-particle trajectories, and applies in general to any indistinguishable boson particles, including massive ones. Recently, this effect has been observed in optical tweezers [29] and atomic beams with Bose-Einstein condensates [30].

Our experimental apparatus is ideally suited to implement a direct analog of the original HOM experiment, thus demonstrating the essential building block to study correlated discrete-time quantum walks with indistinguishable particles.

Continuous-time analogues of DTQWs with correlated boson particles have simi-larly been observed [31, 32].

In our experimental realization of the atomic HOM effect, we initially prepare two atoms separated by 20 lattice sites, cool them into the 3D ground state, and transport them into the same lattice site using an adiabatic ramp (see Fig. 10(a,b)). Instead of letting atoms interact on a millisecond time scale (see Sec. 6), we directly apply a microwave $\pi/2$ pulse (see Fig. 11(a)) rotating the spin of both atoms onto the equator of the Bloch sphere in a time (5 μs) much shorter than any other time scale. By subsequently displacing our sublattices spin dependently, we expect to produce the entangled NOON state $(|\text{Right}, \uparrow\uparrow\rangle - |\text{Left}, \downarrow\downarrow\rangle)/\sqrt{2}$, as shown in Fig. 11(b). Analogously to the original detection method [28], we provide experimental signa-ture of the two-particle interference by recording the suppression of events where two atoms are detected at different lattice sites, corresponding to an anti-bunched state. Ideally these events should not occur, yet they are detected in the experiment due to imperfect ground state cooling, which impairs the indistinguishability. In ad-dition, our detection method does not allow a direct, unambiguous identification of all physical events due to technical reasons (finite survival probability and imperfect efficiency of parity projection [32]). Therefore, we need to resort to a Monte Carlo analysis of our results relying on experimental parameters measured independently. Our analysis yields a signature of the HOM interference with a statical significance of about 3σ. In addition, the Monte Carlo analysis confirms that the observed sup-pression of events with anti-bunched atoms is compatible with a 3D ground state population of around 60%. A detailed investigation of systematic effects affecting our Monte Carlo analysis will be the subject of future work.

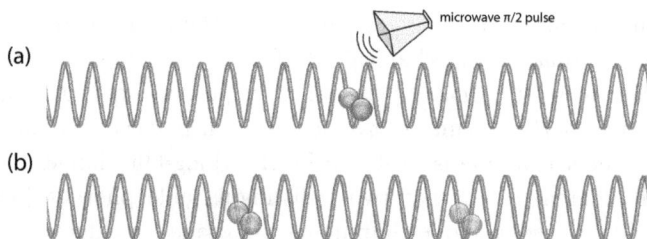

Fig. 11. Scheme of the microwave Hong-Ou-Mandel interferometer with massive particles. (a) Two atoms with opposite spin states are transported to the same lattice site in an analogous way as in Fig. 10. A microwave $\pi/2$ pulse mixes the two indistinguishable atoms like in the famous optical realization of the two-photon interferometer. (b) After the two spin species are separated, both atoms emerge either on the left or right hand side. For identical atoms, no event is expected with the two atoms in distinct sites.

8. Conclusions

We reviewed the state of the art of quantum walk experiments in state-dependent optical lattices. The ability to transport atoms in a spin dependent fashion through two fully independent sublattice potentials opens new ways to study the physics of two or few quantum particles. One long-term goal consists in realizing quantum cellular automata of interacting, indistinguishable particles.

The challenge for the future is to extend this technology from one-dimensional to two-dimensional lattices. This will allow us to study, in particular, new topological phases, where atomic matter waves propagating in a unidirectional fashion along the boundary of a topological island are expected to manifest [33].

Acknowledgements

The authors gratefully thank Wolfgang Alt and Jean-Michel Raimond for insightful discussions, and Gautam Ramola for helpful contributions. The authors also acknowledge financial support from NRW-Nachwuchsforschergruppe "Quantenkontrolle auf der Nanoskala", ERC grant DQSIM, EU project SIQS. In addition, CR and SB from BCGS program and CR from Studienstiftung des deutschen Volkes.

References

[1] M. Raizen, C. Salomon and Q. Niu, New light on quantum transport, *Phys. Today* **50**, p. 30 (1997).
[2] A. Alberti, V. V. Ivanov, G. M. Tino and G. Ferrari, Engineering the quantum transport of atomic wavefunctions over macroscopic distances, *Nat. Phys.* **5**, p. 547 (2009).
[3] A. M. Childs, Universal computation by quantum walk, *Phys. Rev. Lett.* **102**, p. 180501 (2009).
[4] N. Shenvi, J. Kempe and K. B. Whaley, Quantum random-walk search algorithm, *Phys. Rev. A* **67**, p. 052307 (2003).
[5] M. Karski, L. Förster, J. Choi, A. Steffen, W. Alt, D. Meschede and A. Widera, Quantum walk in position space with single optically trapped atoms, *Science* **325**, p. 174 (2009).
[6] F. Zähringer, G. Kirchmair, R. Gerritsma, E. Solano, R. Blatt and C. F. Roos, Realization of a quantum walk with one and two trapped ions, *Phys. Rev. Lett.* **104**, p. 100503 (2010).
[7] R. Matjeschk, C. Schneider, M. Enderlein, T. Huber, H. Schmitz, J. Glueckert and T. Schaetz, Experimental simulation and limitations of quantum walks with trapped ions, *New J. Phys.* **14**, p. 035012 (2012).
[8] M. A. Broome, A. Fedrizzi, B. P. Lanyon, I. Kassal, A. Aspuru-Guzik and A. G. White, Discrete single-photon quantum walks with tunable decoherence, *Phys. Rev. Lett.* **104**, p. 153602 (2010).
[9] A. Schreiber, K. N. Cassemiro, V. Potoček, A. Gábris, P. J. Mosley, E. An-

dersson, I. Jex and C. Silberhorn, Photons walking the line: A quantum walk with adjustable coin operations, *Phys. Rev. Lett.* **104**, p. 050502 (2010).

[10] L. Sansoni, F. Sciarrino, G. Vallone, P. Mataloni, A. Crespi, R. Ramponi and R. Osellame, Two-particle bosonic-fermionic quantum walk via integrated photonics, *Phys. Rev. Lett.* **108**, p. 010502 (2012).

[11] M. Karski, L. Förster, J. Choi, W. Alt, A. Widera and D. Meschede, Nearest-neighbor detection of atoms in a 1d optical lattice by fluorescence imaging, *Phys. Rev. Lett.* **102**, p. 053001 (2009).

[12] S. Kuhr, W. Alt, D. Schrader, I. Dotsenko, Y. Miroshnychenko, A. Rauschenbeutel and D. Meschede, Analysis of dephasing mechanisms in a standing-wave dipole trap, *Phys. Rev. A* **72**, p. 023406 (2005).

[13] D. Jaksch, H.-J. Briegel, J. I. Cirac, C. W. Gardiner and P. Zoller, Entanglement of atoms via cold controlled collisions, *Phys. Rev. Lett.* **82**, p. 1975 (1999).

[14] M. Genske, W. Alt, A. Steffen, A. H. Werner, R. F. Werner, D. Meschede and A. Alberti, Electric quantum walks with individual atoms, *Phys. Rev. Lett.* **110**, p. 190601 (2013).

[15] C. Robens, W. Alt, D. Meschede, C. Emary and A. Alberti, Ideal negative measurements in quantum walks disprove theories based on classical trajectories, *Phys. Rev. X* **5**, p. 011003 (2015).

[16] A. Steffen, A. Alberti, W. Alt, N. Belmechri, S. Hild, M. Karski, A. Widera and D. Meschede, A digital atom interferometer with single particle control on a discretized spacetime geometry, *Proc. Natl. Acad. Sci. U.S.A* **109**, p. 9770 (2012).

[17] A. Alberti, W. Alt, R. Werner and D. Meschede, Decoherence models for discrete-time quantum walks and their application to neutral atom experiments, *New J. Phys.* **16**, p. 123052 (2014).

[18] C. Cedzich, T. Rybár, A. H. Werner, A. Alberti, M. Genske and R. F. Werner, Propagation of quantum walks in electric fields, http://dx.doi.org/10.1103/PhysRevLett.111.160601, *Phys. Rev. Lett.* **111**, p. 160601 (2013).

[19] R. P. Feynman and A. R. Hibbs, *Quantum Mechanics and Path Integrals* (McGraw-Hill, New York, 1965).

[20] A. Bassi, K. Lochan, S. Satin, T. P. Singh and H. Ulbricht, Models of wave-function collapse, underlying theories, and experimental tests, http://dx.doi.org/10.1103/RevModPhys.85.471, *Rev. Mod. Phys.* **85**, p. 471 (2013).

[21] S. Nimmrichter and K. Hornberger, Macroscopicity of mechanical quantum superposition states, http://dx.doi.org/10.1103/PhysRevLett.110.160403, *Phys. Rev. Lett.* **110**, p. 160403 (2013).

[22] A. J. Leggett and A. Garg, Quantum mechanics versus macroscopic realism: Is the flux there when nobody looks?, *Phys. Rev. Lett.* **54**, p. 857 (1985).

[23] A. Ahlbrecht, A. Alberti, D. Meschede, V. B. Scholz, A. H. Werner and R. F. Werner, Molecular binding in interacting quantum walks, *New J. Phys.* **14**, p. 073050 (2012).

[24] C. Monroe, D. M. Meekhof, B. E. King, S. R. Jefferts, W. M. Itano, D. J. Wineland and P. Gould, Resolved-sideband raman cooling of a bound atom to the 3d zero-point energy, *Phys. Rev. Lett.* **75**, p. 4011 (1995).

[25] L. Förster, M. Karski, J. Choi, A. Steffen, W. Alt, D. Meschede, A. Widera, E. Montano, J. H. Lee, W. Rakreungdet and P. S. Jessen, Microwave control of atomic motion in optical lattices, *Phys. Rev. Lett.* **103**, p. 233001 (2009).

[26] N. Belmechri, L. Förster, W. Alt, A. Widera, D. Meschede and A. Alberti, Microwave control of atomic motional states in a spin-dependent optical lattice, *J. Phys. B: At. Mol. Phys.* **46**, p. 104006 (2013).

[27] J. Söding, D. Guéry-Odelin, P. Desbiolles, G. Ferrari and J. Dalibard, Giant spin relaxation of an ultracold cesium gas, *Phys. Rev. Lett.* **80**, p. 1869 (1998).

[28] C. K. Hong, Z. Y. Ou and L. Mandel, Measurement of subpicosecond time intervals between two photons by interference, *Phys. Rev. Lett.* **59**, p. 2044 (1987).

[29] A. M. Kaufman, B. J. Lester, C. M. Reynolds, M. L. Wall, M. Foss-Feig, K. R. A. Hazzard, A. M. Rey and C. A. Regal, Two-particle quantum interference in tunnel-coupled optical tweezers, *Science* **345**, p. 306 (2014).

[30] R. Lopes, A. Imanaliev, A. Aspect, M. Cheneau, D. Boiron and C. I. Westbrook, Atomic Hong–Ou–Mandel experiment, *Nature* **520**, p. 66 (2015).

[31] A. Peruzzo, M. Lobino, J. C. F. Matthews, N. Matsuda, A. Politi, K. Poulios, X.-Q. Zhou, Y. Lahini, N. Ismail, K. Wörhoff, Y. Bromberg, Y. Silberberg, M. G. Thompson and J. L. O'Brien, Quantum walks of correlated photons, *Science* **329**, p. 1500 (2010).

[32] P. M. Preiss, R. Ma, M. E. Tai, A. Lukin, M. Rispoli, P. Zupancic, Y. Lahini, R. Islam and M. Greiner, Strongly correlated quantum walks in optical lattices, *Science* **347**, p. 1229 (2015).

[33] J. K. Asbóth and J. M. Edge, Edge-state-enhanced transport in a two-dimensional quantum walk, *Phys. Rev. A* **91**, p. 022324 (2015).

Muonic Atoms and the Nuclear Structure

A. Antognini* for the CREMA collaboration

Institute for Particle Physics, ETH, 8093 Zurich, Switzerland
Laboratory for Particle Physics, Paul Scherrer Institute, 5232 Villigen-PSI, Switzerland
* E-mail: aldo.antognini@psi.ch

High-precision laser spectroscopy of atomic energy levels enables the measurement of nuclear properties. Sensitivity to these properties is particularly enhanced in muonic atoms which are bound systems of a muon and a nucleus. Exemplary is the measurement of the proton charge radius from muonic hydrogen performed by the CREMA collaboration which resulted in an order of magnitude more precise charge radius as extracted from other methods but at a variance of 7 standard deviations. Here, we summarize the role of muonic atoms for the extraction of nuclear charge radii, we present the status of the so called "proton charge radius puzzle", and we sketch how muonic atoms can be used to infer also the magnetic nuclear radii, demonstrating again an interesting interplay between atomic and particle/nuclear physics.

Keywords: Proton radius; Muon; Laser spectroscopy, Muonic atoms; Charge and magnetic radii; Hydrogen; Electron-proton scattering; Hyperfine splitting; Nuclear models.

1. What atomic physics can do for nuclear physics

The theory of the energy levels for few electrons systems, which is based on bound-state QED, has an exceptional predictive power that can be systematically improved due to the perturbative nature of the theory itself [1, 2]. On the other side, laser spectroscopy yields spacing between energy levels in these atomic systems so precisely, that even tiny effects related with the nuclear structure already influence several significant digits of these measurements. Thus, highly accurate atomic transition frequency measurements can be used as precise and clean probes (purely electromagnetic interaction) of low energy-QCD properties of the nucleus due to the low energy nature of the photons articulating the interaction between the nucleus and the orbiting particle.

A particular class of atoms, called muonic atoms, offer an interesting opportunity to extract properties of the nucleus with high accuracy. In these atoms, one or more electrons are substituted by a muon, which is a fundamental particle having the same electromagnetic properties as the electron but with a much larger mass ($m_\mu \approx 200 m_e$). For example muonic hydrogen (μp) is the bound system of a negative muon and a proton, muonic helium ion (μHe^+) a muon bound to an alpha particle. The atomic properties are strongly affected by the orbiting particle mass m, e.g., the Bohr energy scales linearly with m while the Bohr radius as $1/m$, resulting already for low-Z atoms in muonic binding energies of several keV and in a so strong overlap of the muon wave functions with the nucleus that the energy levels are considerably (%-level) affected by the nucleus finite size. A paradigmatic example is μp whose laser spectroscopy yielded a very precise determination of the proton charge radius [3, 4].

2. Charge and magnetic radii of the proton from scattering

The scattering process between charged particles without internal structure, as for example electron-electron scattering can be fully described within QED. Oppositely, when describing electron-proton scattering, form factors need to be introduced to parametrize the complexity of the nuclear structure. They contain dynamical information on the electric and magnetic currents in the nucleus defining the response to the electromagnetic fields. As a consequence of current conservation and relativistic invariance, for the spin-1/2 nuclei, as protons, only two form factors are required. Experimentally these form factors can be accessed through measurements of the elastic differential cross section which in the one-photon approximation is [5]

$$\left(\frac{d\sigma}{d\Omega}\right)_{\text{elastic}} = \left(\frac{d\sigma}{d\Omega}\right)_{\text{Mott}} \times \frac{1}{1+\tau}\left(G_E^2(Q^2) + \frac{\tau}{\varepsilon}G_M^2(Q^2)\right), \tag{1}$$

where the Mott cross section applies for point-like particles and is fully calculated in the QED framework. $G_E(Q^2)$ and $G_M(Q^2)$ are the electric and magnetic Sachs form factors, while $\tau = Q^2/4M^2$ and $\epsilon^{-1} = 1 + 2(1+\tau)\tan^2(\theta/2)$ are kinematical variables with θ being the electron scattering angle and M the nucleus mass. At $Q^2 = 0$ the form factors correspond to the total charge in units of e and magnetic moment in units of the proton magneton: for the proton $G_E^p(0) = 1$ and $G_M^p(0) = 2.793$. So in first approximation at low momentum exchange the response of the nucleus to electromagnetic fields is ruled by its charge and magnetic moment.

Viewed as a Taylor series the charge and the magnetic moment are the first terms in an infinite list of parameters which describes the interaction of the proton with the electromagnetic fields [6]. The next parameters would be the slopes of the electric and magnetic form factors at zero momentum exchange:

$$R_E = -\frac{6}{G_E(0)}\frac{dG_E}{dQ^2}\bigg|_{Q^2=0} \quad \text{and} \quad R_M = -\frac{6}{G_M(0)}\frac{dG_M}{dQ^2}\bigg|_{Q^2=0}. \tag{2}$$

These equations represent the covariant definition of charge and magnetic radii, which in a non-relativistic approximation correspond to the second moments of the electric charge and magnetization distributions ρ_E and ρ_M of the nucleus

$$R_{E/M}^2 \approx \int d\vec{r}\,\rho_{E/M}(\vec{r})r^2. \tag{3}$$

Any hadron/nuclear theory must reproduce these radii being parameters as fundamental as the charge, mass and magnetic moment. Although lattice QCD shows an impressive progress [7], currently these radii can not be accurately predicted from ab-initio theories and their knowledge relies on experiments [5, 8].

The traditional way to extract the form factors from the measured differential cross sections is based on the Rosenbluth separation techniques which consist in plotting the reduced cross section σ_{red} versus ε:

$$\sigma_{\text{red}} \equiv \frac{\varepsilon(1+\tau)}{\tau} \frac{\left(\frac{d\sigma}{d\Omega}\right)_{\text{elastic}}}{\left(\frac{d\sigma}{d\Omega}\right)_{\text{Mott}}} = G_M^2 + \frac{\varepsilon}{\tau}G_E^2. \tag{4}$$

The reduced cross section is linear in ε, with the slope proportional to G_E^2 and the intercept equal to G_M^2. So both form factors can be deduced by measuring $\left(\frac{d\sigma}{d\Omega}\right)_{\text{elastic}}$ at several values of ε which is achieved by varying the electron beam energy and the electron scattering angle while keeping Q^2 fixed.

After this G_E/G_M separation, each measured form factor can be fitted with a polynomial expansion of the form [9]

$$G_{E/M}(Q) = G_{E/M}(0)\left[1 - \frac{Q^2}{6}\langle r_{E/M}^2 \rangle + \frac{Q^4}{120}\langle r_{E/M}^4 \rangle - \ldots\right], \tag{5}$$

where $\langle r_{E/M}^N \rangle$ represent the N-th moments of the charge/magnetic distributions ($\langle r_{E/M}^2 \rangle = R_{E/M}^2$). At very low Q^2, one could hope that the higher moments terms are sufficiently small, such that the $\langle r_{E/M}^2 \rangle$-term can be determined without using a specific model for the form factor. However, at low Q^2 also the $\langle r_{E/M}^2 \rangle$-term becomes increasingly small relative to the first term of the expansion resulting in a loss of sensitivity. So in practice to fit the measured form factors and extract the radii it is necessary to include data at intermediate Q^2. As cross sections data are available only down to a minimal Q^2, and because an extrapolation to $Q^2 = 0$ is required, the choice of the fit function (form factor model) is very important.

This extrapolation is even more challenging for the magnetic radii because of the ε/τ-dependence in Eq. (1) which results in an additional suppression of sensitivity (at low Q^2) of the measured cross sections to G_M compared to G_E. Consequently, the increased uncertainties of G_M at low Q^2 yields magnetic radii with larger uncertainties relative to charge radii. This calls for alternative determinations of the magnetic radii such as from polarized-recoil scattering [5] or atomic spectroscopy.

3. Charge and magnetic radii of the proton from atomic physics

The finite radius of the nucleus implies that its charge is smeared over a finite volume. For hydrogen-like S-states there is a non-negligible probability that the "orbiting" particle is spending some time inside the nuclear charge distribution, thus experiencing a reduced electrostatic attraction as compared to a point-like nucleus. This reduced attraction caused by the modification of the Coulomb potential for very small distances is giving rise to a shift of the atomic energy levels which for S-states H-like systems in leading order reads [1, 2]

$$\Delta E_{\text{finite size}} = \frac{2\pi Z\alpha}{3}|\phi^2(0)|^2 R_E^2 = \frac{2m_r^3(Z\alpha)^4}{3n^3}R_E^2, \tag{6}$$

where $\phi(0)$ is the wave function at the origin in coordinate space, $m_r = mM/(m + M)$ the reduced mass of the atomic system with m being the orbiting particle mass, and M the nucleus mass, α the fine structure constant, Z the charge number of the nucleus and n the principal quantum number.

The m_r^3 dependence of Eq. (6) reveals the advantages related with muonic atoms. As the muon mass is 200 times larger than the electron mass, the muonic wave function strongly overlaps with the nucleus ensuing a large shift of the energy levels due to the nuclear finite size. Thus, the muonic bound-states represent ideal systems for the precise determination of nuclear charge radii R_E [3, 4, 10].

Because of this sensitivity to the finite size a moderate (20 ppm) accuracy in the measurement of the 2S-2P transition in μp is sufficient to extract the proton charge radius very accurately (5×10^{-4} relative accuracy) [3, 4]. In regular H, the accuracies of the transition frequency measurements, also relative to the line-widths, have to be much higher (see Table 1) to compete with this value. By combining in a least-square adjustment all high-precision frequency measurements in H available to date, as accomplished by the CODATA group, a proton charge radius with an accuracy of about 1% is obtained.

Atomic spectroscopy can be used also to extract magnetic radii. This is achieved through precision measurement of hyperfine splittings [4, 11, 12]. For H-like systems, in leading approximation the HFS is given by the magnetic interaction between the nucleus $\vec{\mu}_N$ and the orbiting particle $\vec{\mu}_m$ magnetic moments, described by [1, 2]

$$H \sim \vec{\mu}_N \cdot \vec{\mu}_m \, \delta(\vec{r}) \, , \tag{7}$$

which results in an energy splitting of the 1S state given by the Fermi energy

$$E_F = \frac{8}{3} \frac{Z^3 \alpha^4 m_r^3}{mMn^3} \mu_N \, . \tag{8}$$

The finite size correction to this splitting, which is of second order in perturbation theory, is [2, 12]

$$\Delta E_{\text{Zemach}} = -2(Z\alpha) m_r \, E_F \, R_Z \tag{9}$$

where the Zemach radius R_Z is defined as an integral of the charge and magnetic form factors

$$R_Z = -\frac{4}{\pi} \int_0^\infty \frac{dQ}{Q^2} \left(G_E(Q^2) \frac{G_M(Q^2)}{1 + \kappa_p} - 1 \right) , \tag{10}$$

(with κ_p the proton anomalous magnetic moment). In a non-relativistic approximation R_Z can be expressed, by the first moment of the convolution between charge and magnetic distributions $\rho_E(r)$ and $\rho_M(r)$ in coordinate space

$$R_Z = \int d^3\mathbf{r} \, |\mathbf{r}| \int d^3\mathbf{r}' \rho_E(\mathbf{r} - \mathbf{r}')\rho_M(\mathbf{r}'). \tag{11}$$

When assuming form factor models or using measured form factor data, the magnetic radius can be extracted from the Zemach radius. Thus, accurate measurements of the HFS in μp and H can be used as complementary ways to obtain a precise value of the proton magnetic radius, or alternatively, as presented in [13] a self-consistent value of $R_E^2 + R_M^2$.

4. The proton charge radius puzzle

Three complementary routes to the proton charge radius have been undertaken: the historical method relies on elastic electron-proton scattering, the second one on high-precision laser spectroscopy in H, and the third one on high sensitivity laser spectroscopy in μp. The value extracted from μp [3, 4] with a relative accuracy of 5×10^{-4} is an order of magnitude more accurate than obtained from the other methods. Yet the value is 4% smaller than derived from electron-proton scattering [8, 14, 15] and H spectroscopy [17] with a disagreement at the 7σ level.

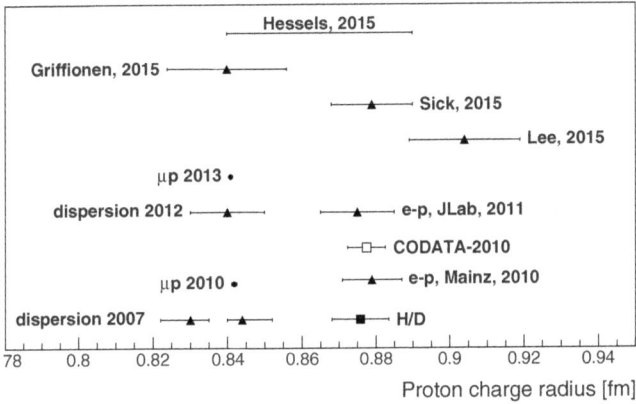

Fig. 1. Proton charge radii determined from spectroscopy of muonic atoms (full circles), from electron scattering (triangles) and from H/D spectroscopy (full squares).

The most recent evaluations of the proton charge radius are summarized in Fig. 1. The most precise values are extracted from two transition frequency measurements in μp. By combining them we obtained a 2S-2P$_{1/2}$ splitting of $\Delta E_{2S-2P_{1/2}}^{\text{exp}} = 202.3706(23)$ meV equivalent to a frequency of 48932.99(55) GHz, limited by statistics while the systematic effects are at the 300 MHz level [4]. Equating this experimental value with the theoretical prediction

$$E_L^{\text{th}} = 206.0336(15) \, [\text{meV}] - 5.2275(10) \left[\frac{\text{meV}}{\text{fm}^2} \right] R_E^2 + 0.0332(20) \, [\text{meV}] \qquad (12)$$

yields the proton charge radius R_E in fm. The first term of Eq. (12) accounts for QED contributions, the second one for finite size effects, and the third one for the two-photon exchange (TPE) contribution which is a second-order perturbation theory contribution related with the proton structure. In the last years as summarized in [10, 18, 19] various cross checks and refinements of bound-state QED and TPE calculations needed for the extraction of R_E from μp have been performed, but no substantial missing effects have been found that could explain the discrepancy.

The typical systematics affecting the atomic energy levels are substantially suppressed in μp due to the stronger binding. The internal fields and the level separation of the muonic atoms are greatly enhanced compared to regular atoms making them insensitive to external fields (AC and DC Stark, Zeeman, black-body and pressure shifts). Thus μp turns out to be very sensitive to the proton charge radius (m_r^3-dependence) and insensitive to systematics which typically scale as $\sim 1/m_r$.

Special attention was devoted to the analysis of electron-proton scattering data and the issues related with the extrapolation procedure. Starting from fit functions given by truncated general series expansions such as Taylor, splines and polynomials a large progress has been achieved in the last years by the use of various techniques: enforcing analyticity [6, 15], constraining the low Q^2 behavior of the form factor assuming a large-r behavior of the charge distribution [14] or by using proton models [20]. Tension exists between various electron-proton data analysis: some give results compatible with μp [20–22], some at variance [8, 14–16]. Because data at even lower Q^2 would facilitate the extrapolation at $Q^2 = 0$, two electron-proton experiments have been initiated, one at JLAB [23], the other one at MAMI Mainz [24]. A comparison between muon-proton and electron-proton scattering within the same setup as proposed by the MUSE [25] collaboration at PSI could disclose a possible violation of muon-electron universality.

Several "beyond standard model" BSM extensions have been studied but the majority of them have difficulties to resolve the discrepancy without conflicting with other low energy constraints. Still some BSM theories can be formulated but they require fine-tuning (e.g. cancellation between axial and vector components), targeted coupling (e.g. preferentially to muons) and are problematic to be merged in a gauge invariant way into the standard model [26, 27]. Breakdown of the perturbative approach in the electron-proton interaction at short distances, as well as the interaction with sea $\mu^+\mu^-$ and e^+e^- pairs and unusual proton structure have been suggested as possible explanation but without conclusive quantification [28] .

Summarizing, currently the discrepancy persists even though recent reanalysis of scattering data have led to larger uncertainties of the extracted proton radius. New data from muonic deuterium and helium, from H spectroscopy and electron-proton scattering holds the potential to clarify the situation in the near future.

5. The proton radius from H spectroscopy

In a simplified way, the hydrogen S-state energy levels can be described by

$$E(nS) = \frac{R_\infty}{n^2} + \frac{L_{1S}}{n^3} , \qquad (13)$$

where $R_\infty = 3.289\,841\,960\,355(19) \times 10^{15}$ Hz is the Rydberg constant and

$$L_{1S} \simeq 8171.636(4) \text{ [MHz]} + 1.5645 \left[\frac{\text{MHz}}{\text{fm}^2}\right] R_E^2 \qquad (14)$$

the 1S Lamb shift given by bound-state QED contributions. The different n-dependence of the two terms in Eq. (13) permits to extract both R_∞ and L_{1S} (thus R_E) from at least two frequency measurements in H.

Being the most precisely known transition (relative accuracy of 4×10^{-15}) [29] and having the largest sensitivity to R_E, usually the 1S-2S transition is used. By combining it with a second transition measurement, R_∞ is eliminated and R_E can be extracted. When taken individually, the various R_E values extracted from H spectroscopy by combining two frequency measurements (2S-4S, 2S-12D, 2S-6S, 2S-6D, 2S-8S, 1S-3S as "second" transition [17]) are statistically compatible with the value from μp. Only the value extracted by pairing the 1S-2S and the 2S-8D transitions is showing a 3σ deviation while all the others differ only by $\lesssim 1.5\sigma$.

So the 4σ discrepancy between the proton charge radius from μp and H spectroscopy emerges only after an averaging process (mean square adjustments of all measured transitions) of the various "individual" determinations and consequently is less startling than it looks at first glance. A small systematic effect common to the H measurements could be sufficient to explain the deviation between μp and H results. This fact becomes even more evident if we consider the frequency shifts (absolute and normalized to the line-width) necessary to match the R_E values from μp and H, as summarized for selected transitions in Table 1. Obviously the discrepancy

Table 1. Relative accuracy of the various transition measurements in H, and hypothetical shift of the measured transition frequencies needed to match the R_E from H and μp. This shift is expressed also relative to the experimental accuracy σ, and to the transition effective line-widths Γ_{eff}.

Transition	Relative accuracy	Shift in σ	Shift in Hz	Shift in line-width
μp(2S-2P)	2×10^{-5}	$100\,\sigma$	75 GHz	$4\,\Gamma_{\text{eff}}$
H(1S-2S)	4×10^{-15}	$4'000\,\sigma$	40 kHz	$40\,\Gamma_{\text{eff}}$
H(2S-4P)	3×10^{-11}	$1.5\,\sigma$	9 kHz	$7 \times 10^{-4}\,\Gamma_{\text{eff}}$
H(2S-2P)	1×10^{-6}	$1.5\,\sigma$	5 kHz	$7 \times 10^{-4}\,\Gamma_{\text{eff}}$
H(2S-8D)	9×10^{-12}	$3\,\sigma$	20 kHz	$2 \times 10^{-2}\,\Gamma_{\text{eff}}$
H(2S-12D)	1×10^{-11}	$1\,\sigma$	8 kHz	$5 \times 10^{-3}\,\Gamma_{\text{eff}}$
H(1S-3S)	4×10^{-12}	$1\,\sigma$	13 kHz	$5 \times 10^{-3}\,\Gamma_{\text{eff}}$

can not be solved by slightly tuning (shifting) the measured values of the 1S-2S transition in H and the 2S-2P transitions in μp because it would require displacements corresponding to $4000\,\sigma$ and $100\,\sigma$, respectively. Expressing the required frequency shift relative to the line-width as in the last column allows to better recognize some aspects of the experimental challenges. For example a shift of only $7 \times 10^{-4}\,\Gamma$ of the 2S-4P transition would be sufficient to explain the discrepancy. A control of the systematics which could distort and shift the line shape on this level of accuracy is far from being a trivial task. Well investigated are the large line broadening owing to inhomogeneous light shifts which results in profiles with effective widths much larger than the natural line-widths [17].

Another exemplary correction relevant in this context, named quantum interference, has been brought recently back to attention [30], and has lead to various reevaluations of precision experiments. An atomic transition can be shifted by the presence of a neighboring line, and this energy shift δE, as a rule of thumb, amounts maximally to $\frac{\delta E}{\Gamma} \approx \frac{\Gamma}{D}$ where D is the energy difference between the two resonances and Γ the transition line-width. Thus, if a transition frequency is aimed with an absolute accuracy of Γ/x, then the influence of the neighboring lines with $D \leq x\Gamma$ has to be considered. The precise evaluation of these quantum interference effects are challenging because they require solving numerous differential equations describing the amplitude of the total excitation and detection processes from initial to final state distributions which depends on the details of the experimental setup.

Generally speaking, transition frequencies involving states with large n are more sensitive to systematic effects caused by external fields. Emblematic is the n^7-dependence of the Stark effect. Motivated by the possibility that minor effects in H could be responsible for the discrepancy, various activities have been initiated in this field: at MPQ Garching the 2S-4P [31] and 1S-3S transitions are addressed, at LKB Paris the 1S-3S [32], and at the Toronto university the 2S-2P [33].

The "second" (beside the 1S-2S transition) transition frequency measurement in H can be interpreted as a R_∞ determination. Optical spectroscopy of H-like ions between circular Rydberg states where the nuclear size corrections are basically absent, the QED contributions small, and the line-widths narrow can be used as alternative determination of R_∞ [34]. Another way to R_∞ is through spectroscopy of muonium and positronium atoms which are purely leptonic systems where uncertainties related with the finite size are absent [35].

6. Hyperfine splitting in μp and μ^3He$^+$

As a next step, we plan to prepare the measurement by means of laser spectroscopy of the ground state hyperfine splitting (1S-HFS) in μp and μ^3He$^+$ with few ppm relative accuracy. Similar activities in μp exist at RIKEN-RAL and J-PARC [36, 37]. The theoretical prediction for the 1S-HFS in μp is approximately [11, 12, 38, 39]

$$\Delta E_{\text{HFS}}^{\text{th}} = 182.819(1) \text{ [meV]} - 1.301 \left[\frac{\text{meV}}{\text{fm}}\right] R_Z + 0.064(21) \text{ [meV]} , \qquad (15)$$

where the first term includes the Fermi energy, QED corrections, hadronic vacuum polarization, recoil corrections and weak interactions. These contributions are known well enough. The second term is the finite size contribution, which is proportional to R_Z. It contains also some higher order mixed radiative finite-size corrections. The third term is given by the proton polarizability contribution.

By comparing the theoretical prediction with the experiment, it will become possible to deduce R_Z with a relative accuracy better than 5×10^{-3} provided that the polarizability contribution will be improved below 10% relative accuracy. This contribution can be computed using a dispersive approach and measured proton polarized structure function g_1 and g_2 [38, 39] or via chiral perturbation theories (ChPT) [40]. An improvement of this contributions is conceivable in the near future due to the considerable advance in ChPT [41] and due to various ongoing measurements of the proton structure functions at JLAB using polarized target and beams.

For μ^3He^+ the situation is conceptually similar to μp. The theoretical predictions assumes the same form as in Eq. (15) but with different numerical values.

The motivations for these experiments are several:

- **Bound-state QED in H and understanding of the 21 cm line**

  The uncertainty of R_Z presently limits, together with the polarizability contribution, the theoretical prediction of the 1S-HFS in H. Therefore, the comparison between the experimental 1S-HFS value in H, which has a relative accuracy smaller than 10^{-12}, with the theoretical predictions is limited by the uncertainty of the proton structure contributions. This situation can be improved by complementary measurements in μp opening the way for a test of the HFS in H at the 10^{-7} level of accuracy.

- **Understanding of the proton structure**

  Practically, from the Zemach radius the magnetic radius can be obtained by using form factor models or measured form factor data. As the determination of R_M from elastic electron-proton scattering is very challenging due to the loss of sensitivity for the magnetic form factor with decreasing momentum exchange, a precise measurement of R_Z from the muonic HFS represents a valuable complementary route to R_M. It can be used also to sort out a 8% discrepancy between R_M as extracted from the recent unpolarized electron-proton cross sections measurements in Mainz, and as deduced from polarized-recoil data at JLAB [6, 8, 15, 20].

  Currently, we cannot determine the radii and the form factors accurately from theory, although lattice QCD is making impressive progress on this issue [7]. A precise measurement of R_Z from μp and its comparison with correlative measurements from scattering experiments bears the potential to push the frontier of our understanding of the complex non-perturbative nature of the proton structure which has been deeply reviewed in the last 15 years especially due to polarization data and the development of theoretical tools such as chiral perturbation theory.

  The interplay between the muonic measurement and investigations of the proton structure can be articulated in several ways. As mentioned previously R_Z (R_M) represents a benchmark for the understanding of the proton structure. Extraction of a precise value for the Zemach radius from the μp 1S-HFS measurement requires the knowledge of the proton polarizability contribution which requires modeling of the proton and data from scattering (ChPT, g_1 and g_2 structure functions). Inverting this logic, a precise value of R_Z from scattering data [42] can be thus used, when paired with the μp HFS, to check the polarizability contribution.

- **Nuclear physics from $\mu^3\mathrm{He}^+$**

  Nuclei like $^3\mathrm{He}$ are calculable very precisely by a wide variety of ab-initio methods and so provide an important comparison between experiment and theoretical models of both the nuclear interactions (potential) and the electromagnetic currents [43]. The magnetic distribution and magnetic radii turn out to be very sensitive to the meson-exchange currents. A very hot topic in hadronic physics is to measure various parton distributions to see the quark spin distribution within protons and neutrons. The same should be done for nucleon spin distributions in nuclei.

7. Conclusions

Precision measurements in muonic atoms have triggered a plethora of theoretical works and experimental investigations in various fields of physics showing the potential and interdisciplinarity of these precision experiments [27]. Spectroscopy of the 2S-2P splittings in μp, μd, $\mu^3\mathrm{He}^+$ and $\mu^4\mathrm{He}^+$ has been accomplished by the CREMA collaboration. Besides the proton charge radius, soon new accurate values of the deuteron and $^3\mathrm{He}$ and $^4\mathrm{He}$ nuclear radii will be extracted from these measurements providing insights into the proton radius puzzle, and benchmarks to check few-nucleon ab-initio calculations. Moreover they can be used as anchor point for the $^6\mathrm{He}$-$^4\mathrm{He}$ and $^8\mathrm{He}$-$^4\mathrm{He}$ isotopic shift measurements [44] and their knowledge opens the way to enhanced bound-state QED tests for one- and two-electrons systems in "regular" He^+ [45] and He [46].

Spectroscopy of HFS transitions in μp and $\mu^3\mathrm{He}^+$ provides a natural continuation of the CREMA program. Letting aside the proton radius puzzle related "new physics" searches the 1S-HFS in μp and $\mu^3\mathrm{He}^+$ measurements impact three aspects of fundamental physics: bound-state QED in H-like systems, our understanding of the magnetic distributions and the low-energy spin structure of proton and $^3\mathrm{He}$ nucleus.

Acknowledgments

This work is supported by the Swiss National Science Foundation Projects No. 200021L_138175 and No. 200020_159755. We acknowledge fruitful discussions with N. Barnea, R. Wiringa, I. Sick and V. Pascalutsa.

References

[1] M. I. Eides, H. Grotch and V. A. Shelyuto, Theory of light hydrogenlike atoms, *Phys. Rep.* **342**, 63 (2001).

[2] S. G. Karshenboim, Precision physics of simple atoms: QED tests, nuclear structure and fundamental constants, *Phys. Rep.* **422**, 1 (2005).

[3] R. Pohl, A. Antognini, F. Nez et al., The size of the proton., *Nature* **466**, 213 (2010).

[4] A. Antognini, F. Nez, K. Schuhmann et al., Proton structure from the measurement of 2S-2P transition frequencies of muonic hydrogen, *Science* **339**, 417 (2013).

[5] V. Punjabi, C. F. Perdrisat, M. K. Jones, E. J. Brash and C. E. Carlson, The structure of the nucleon: Elastic electromagnetic form factors, *arXiv*: 1503.01452 (2015).

[6] Z. Epstein, G. Paz and J. Roy, Model independent extraction of the proton magnetic radius from electron scattering, *Phys. Rev. D* **90**, 074027 (2014).

[7] J. R. Green, M. Engelhardt, S. Krieg, J. W. Negele, A. V. Pochinsky and S. N. Syritsyn, Nucleon structure from Lattice QCD using a nearly physical pion mass, *Phys. Rev. D* **90**, 074507 (2014).

[8] J. Bernauer, M. Distler, J. Friedrich et al., Electric and magnetic form factors of the proton, *Phys. Rev. C* **90**, 015206 (2014).

[9] I. Sick, On the rms-radius of the proton, *Phys. Lett. B* **576**, 62 (2003).

[10] R. Pohl, R. Gilman, G. A. Miller and K. Pachucki, Muonic Hydrogen and the Proton Radius Puzzle, *Annu. Rev. Nucl. Part. Sci.* **63**, 175 (2013).

[11] A. V. Volotka, V. M. Shabaev, G. Plunien and G. Soff, Zemach and magnetic radius of the proton from the hyperfine splitting in hydrogen, *Eur. Phys. J. D* **33**, 23 (2005).

[12] A. Dupays, A. Beswick, B. Lepetit, C. Rizzo and D. Bakalov, Proton Zemach radius from measurements of the hyperfine splitting of hydrogen and muonic hydrogen, *Phys. Rev. A* **68**, 052503 (2003).

[13] S. G. Karshenboim, Model-independent determination of the magnetic radius of the proton from spectroscopy of ordinary and muonic hydrogen, *Phys. Rev. D* **90**, 053013 (2014).

[14] I. Sick and D. Trautmann, Proton root-mean-square radii and electron scattering, *Phys. Rev. C* **89**, 012201 (2014).

[15] G. Lee, J. R. Arrington and R. J. Hill, Extraction of the proton radius from electron-proton scattering data, *Phys. Rev. D* **92**, 013013 (2015).

[16] M. O. Distler, T. Walcher and J. C. Bernauer, Solution of the proton radius puzzle? Low momentum transfer electron scattering data are not enough, *arXiv*: 1511.00479 (2015).

[17] B. de Beauvoir, C. Schwob, O. Acef, L. Jozefowski, L. Hilico, F. Nez, L. Julien, A. Clairon and F. Biraben, Metrology of the H and D atoms: Determination of the Rydberg constant and Lamb shifts, *Eur. Phys. J. D* **12**, 61 (2000).

[18] S. Karshenboim et al., Theory of lamb shift in muonic hydrogen, *J. Phys. Chem. Ref. Data* **44**, 031202 (2015).

[19] C. Peset and A. Pineda, The Lamb shift in muonic hydrogen and the proton radius from effective field theories, *Eur. Phys. J. A* **51**, 1-19 (2015).

[20] I. Lorenz, U.-G. Meißner, H.-W. Hammer and Y.-B. Dong, Theoretical constraints and systematic effects in the determination of the proton form factors, *Phys. Rev. D* **91**, 014023 (2015).

[21] D. W. Higinbotham, A. A. Kabir, V. Lin, D. Meekins, B. Norum and B. Sawatzky, The proton radius from electron scattering data, arXiv:1510.01293 (2015).

[22] K. Griffioen, C. Carlson and S. Maddox, Are electron scattering data consistent with a small proton radius?, *Phys. Rev. C* **93**, 065207 (2015).

[23] A. Gasparian, The PRad experiment and the proton radius puzzle, *EPJ Web Conf.* **73**, 07006 (2014).

[24] M. Mihovilovic, H. Merkel and A. Weber, Puzzling out the proton radius puzzle, *EPJ Web of Conf.* **81**, 01009 (2014).

[25] R. Gilman, Studying the proton radius puzzle with μp elastic scattering, *AIP Conf. Proc.* **1563** 167 (2013).

[26] S. G. Karshenboim, D. McKeen and M. Pospelov, Constraints on muon-specific dark forces, *Phys. Rev. D* **90**, 073004 (2014).

[27] C. E. Carlson, The proton radius puzzle, *Prog. Part. Nucl. Phys.* **82**, 59 (2015).

[28] U. D. Jentschura, Muonic bound systems, virtual particles, and proton radius, *Phys. Rev. A* **92**, 012123 (2015).

[29] A. Matveev, C. G. Parthey, K. Predehl, J. Alnis, A. Beyer, R. Holzwarth, T. Udem, T. Wilken, N. Kolachevsky, M. Abgrall, D. Rovera, C. Salomon, P. Laurent, G. Grosche, O. Terra, T. Legero, H. Schnatz, S. Weyers, B. Altschul and T. W. Hänsch, precision measurement of the hydrogen 1S-2S frequency via a 920-km fiber link, *Phys. Rev. Lett.* **110**, 230801 (2013).

[30] M. Horbatsch and E. A. Hessels, Shifts from a distant neighboring resonance, *Phys. Rev. A* **82**, 052519 (2010).

[31] A. Beyer, L. Maisenbacher, K. Khabarova, A. Matveev, R. Pohl, T. Udem, T. W. Hänsch and N. Kolachevsky, Precision spectroscopy of 2S-nP transitions in atomic hydrogen for a new determination of the Rydberg constant and the proton charge radius, *Phys. Scr.* **T165**, 014030 (2015).

[32] S. Galtier, H. Fleurbaey, S. Thomas, L. Julien, F. Biraben and F. Nez, Progress in spectroscopy of the 1s-3s transition in hydrogen, *J. Phys. Chem. Ref. Data* **44**, 031201 (2015).

[33] A. Vutha, N. Bezginov, I. Ferchichi, M. George, V. Isaac, C. Storry, A. Weatherbee, M. Weel and E. Hessels, Progress towards a new microwave measurement of the hydrogen n=2 Lamb shift: A measurement of the proton charge radius, *Bull. Am. Phys. Soc.* **57,** (2012).

[34] J. N. Tan, S. M. Brewer and N. D. Guise, Experimental efforts at NIST towards one-electron ions in circular Rydberg states, *Phys. Scr.* **T144**, 014009 (2011).

[35] D. A. Cooke, P. Crivelli, J. Alnis, A. Antognini, B. Brown, S. Friedreich, A. Gabard, T. W. Hänsch, K. Kirch, A. Rubbia and V. Vrankovic, Observation of positronium annihilation in the 2S state: Towards a new measurement of the 1S-2S transition frequency, *Hyperfine Interact.* **233** 67 (2015).

[36] A. Adamczak, D. Bakalov, L. Stoychev and A. Vacchi, Hyperfine spectroscopy of muonic hydrogen and the PSI Lamb shift experiment, *Nucl. Instr. and Meth. Phys. Res. B* **281**, 72 (2012).

[37] M. Sato, K. Ishida, M. Iwasaki, S. Kanda, Y. Ma, Y. Matsuda, T. Matsuzaki, K. Midorikawa, Y. Oishi, S. Okada, N. Saito, K. Tanaka, Laser spectroscopy of the hyperfine splitting energy in he ground state of muonic hydrogen. 10.3204/DESY-PROC-2014-04/67 (2014).

[38] R. N. Faustov, A. P. Martynenko, G. A. Martynenko and V. V. Sorokin, Radiative nonrecoil nuclear finite size corrections of order $\alpha(Z\alpha)^5$ to the HFS of S-states in μp, *Phys. Lett. B* **733**, 354 (2014).

[39] C. E. Carlson, V. Nazaryan and K. Griffioen, Proton-structure corrections to hyperfine splitting in muonic hydrogen, *Phys. Rev. A* **83**, 042509 (2011).

[40] F. Hagelstein and V. Pascalutsa, Proton structure in the hyperfine splitting of muonic hydrogen, *arXiv:1511.04301* (2015).

[41] F. Hagelstein, R. Miskimen and V. Pascalutsa, Nucleon polarizabilities: From compton scattering to hydrogen atom, *Progress in Particle and Nuclear Physics* **88**, 29-97 (2016).

[42] M. O. Distler, J. C. Bernauer and T. Walcher, The RMS charge radius of the proton and Zemach moments, *Phys. Lett. B* **696**, 343 (2011).

[43] I. Sick, Zemach moments of He^3 and He^4, *Phys. Rev. C* **90**, 064002 (2014).

[44] Z.-T. Lu, P. Mueller, G. W. F. Drake, W. Nörtershäuser, S. C. Pieper and Z.-C. Yan, Laser probing of neutron-rich nuclei in light atoms, *Rev. Mod. Phys.* **85**, 1383 (2013).

[45] M. Herrmann, M. Haas, U. D. Jentschura, F. Kottmann, D. Leibfried, G. Saathoff, C. Gohle, A. Ozawa, V. Batteiger, S. Knünz, N. Kolachevsky, H. A. Schüssler, T. W. Hänsch and T. Udem, Feasibility of coherent XUV spectroscopy on the 1S-2S transition in singly ionized helium, *Phys. Rev. A* **79**, 052505 (2009).

[46] D. Z. Kandula, C. Gohle, T. J. Pinkert, W. Ubachs and K. S. E. Eikema, XUV frequency-comb metrology on the ground state of helium, *Phys. Rev. A* **84**, 062512 (2011).

Doppler-Broadening Gas Thermometry at 1.39 μm: Towards a New Spectroscopic Determination of the Boltzmann Constant

A. Castrillo, M.D. De Vizia, E. Fasci, T. Odintsova, L. Moretti and L. Gianfrani*

*Department of Mathematics and Physics, Second University of Naples,
Caserta, I-81100, Italy*
* E-mail: livio.gianfrani@unina2.it
http://www.matfis.unina2.it

The expression of the Doppler width of a spectral line, valid for a gaseous sample at thermodynamic equilibrium, represents a powerful tool to link the thermodynamic temperature to an optical frequency. This is the basis of a relatively new method of primary gas thermometry, known as Doppler broadening thermometry. Implemented at the Second University of Naples on $H_2{}^{18}O$ molecules at the temperature of the triple point of water, this method has recently allowed to determine the Boltzmann constant with a global uncertainty of 24 parts over 10^6. Even though this is the best result ever obtained by using an optical method, its uncertainty is still far from the requirement for the new definition of the unit kelvin. To this end, Doppler broadening thermometry should approach the accuracy of 1 part per million. In this paper, we will report on our recent efforts to further develop and optimize Doppler broadening thermometry at 1.39 μm, using acetylene as a molecular target. Main progresses and current limitations will be highlighted.

Keywords: Spectral line shape; Doppler effect; Boltzmann constant; laser spectroscopy.

1. Introduction

In November 2014, at its 25th meeting, the General Conference for Weights and Measurements (CGPM) adopted a Resolution on the future revision of the International System of Units (SI). In the new SI four base units (namely, the kilogram, the ampere, the kelvin, and the mole) will be redefined in terms of fixed numerical values for a set of fundamental constants (the Planck constant, the elementary charge, the Boltzmann constant, and the Avogadro constant, respectively). This is done to eliminate any artefact or material dependency and ensure the long term stability of the units. In the same Resolution, the CGPM encouraged scientists to keep working on these constants to obtain refined data with the required uncertainties.

The most accurate way to access the value of the Boltzmann constant (k_B) is from measurements of the speed of sound in a noble gas inside an acoustic resonator [1, 2]. After many decades of research and technical developments, acoustic gas thermometry has recently provided a k_B determination with a relative uncertainty of 0.71 parts over 10^6 [3]. Another consolidated approach, based upon the Clausius-Mossotti equation, deals with measurements of the electric susceptibility of helium as a function of the gas pressure. Dielectric constant gas thermometry has recently led to a k_B value with a combined uncertainty of 4.3×10^{-6} [4]. Other upcoming primary thermometry techniques are currently at the stage of further development and optimization. Among them, the newest and most promising one is Doppler broadening thermometry (DBT). As conceived by Christian J. Bordé, DBT consists in retrieving the Doppler width ($\Delta\nu_D$) from the highly accurate observation of the spectral profile corresponding to a given atomic or molecular line in a gas sample at thermodynamic equilibrium [5, 6]. If implemented at the temperature of the triple point of water, under the linear regime of laser-gas interaction, DBT can provide an optical determination of k_B by inverting the following equation:

$$\Delta\nu_D = \frac{\nu_0}{c}\sqrt{2ln2\frac{k_B T}{m}}, \tag{1}$$

where ν_0 is the line center frequency, c is the speed of light, T the thermodynamic temperature, and m the molecular mass.

So far, the most accurate spectroscopic determination of k_B has been reported by our group, using a dual-laser water spectrometer at 1.39 μm [7]. By probing the $4_{4,1}-> 4_{4,0}$ line of the $H_2{}^{18}O$ $\nu_1 + \nu_3$ band, after the application of a sophisticated and extremely refined spectral analysis procedure for the retrieval of the Doppler width as a function of the gas pressure, we found $k_B = 1.380631(33) \times 10^{-23} JK^{-1}$ [7]. This value presents a combined (type A and type B) uncertainty of 24 parts over 10^6, which improves the previous result obtained on CO_2 at 2-μm wavelength by a factor of 6.7 [8]. The complete uncertainty budget was first illustrated in Ref. 9 and then further refined, as reported in Ref. 10. The major sources of uncertainty are due to: i) the statistical fluctuation of the data points arising from individual fits of repeated spectra; ii) the spectral purity of reference and probe lasers; iii) the line shape model.

In this paper, we report on the first results of the third-generation experiment that is being performed at the Second University of Naples, expressly conceived to significantly reduce the three main sources of uncertainty. The molecular target is acetylene (C_2H_2), which shows interesting absorption features at 1.39 μm [11]. This choice offers some important advantages, as compared to water, C_2H_2 being a non-polar and linear molecule. In particular, the fact that the molecule does not present a permanent dipole moment reduces significantly the interactions with the walls of the gas container. Moreover, despite the relatively large number of vibrational modes (five in total, namely, the symmetric C-H stretch, the symmetric C-C stretch, the antisymmetric C-H stretch, the antisymmetric bend, and the symmetric bend), it is possible to find well isolated transitions, such as the P(14) component of the $2\nu_3 + \nu_5^1$ band. The main drawback can be ascribed to the linestrength, which is a factor of 400 smaller than that of the $H_2{}^{18}O$ line.

2. Experimental set-up

The new spectrometer is schematically shown in Figure 1. It basically consists of an extended-cavity diode laser (namely, the probe laser) with an emission wavelength

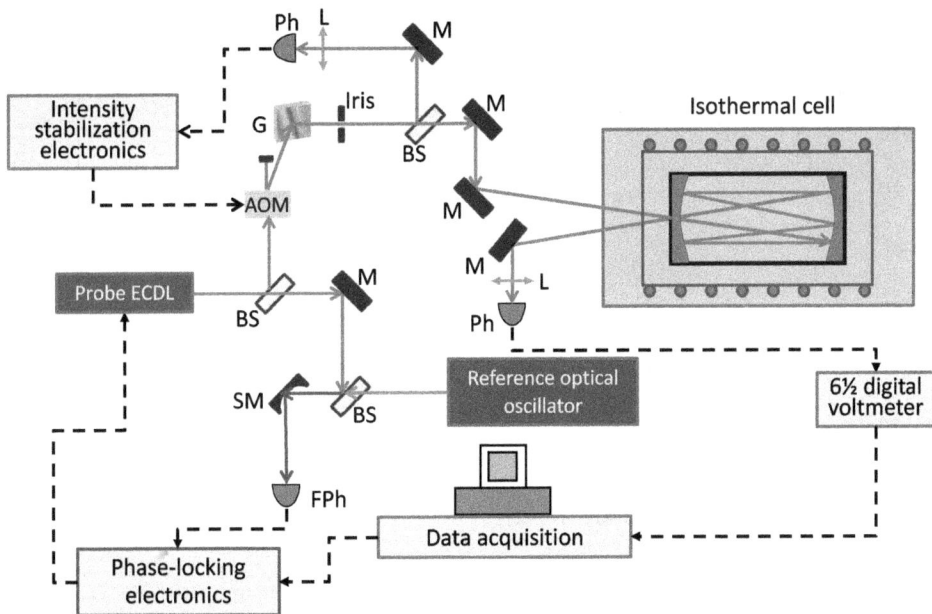

Fig. 1. Sketch of the dual-laser spectrometer. BS: beam splitter; M: mirror; SM: spherical mirror; G: grating; L: anti-reflection coated lens; AOM: acousto-optic modulator; Ph: InGaAs photodiode; FPh: fast photodiode.

in the range between 1.38 and 1.41 μm, a frequency control unit, an intensity stabilization feed-back loop, and an isothermal cell. Precise stabilization and synchro-
nization of the laser frequency is achieved by using the technique of phase-locking,
in which the probe laser is forced to maintain a precise frequency-offset from a reference laser. This offset is provided by a radio-frequency (rf) synthesizer, which
in turn is phase-locked to an ultra-stable Rb oscillator. The optical phase-locking
loop reduced the width of the beat-note between the probe and the reference lasers
to the Hz level, thus demonstrating the narrowing of the probe laser down to the
limit determined by the spectral purity of the reference laser. An example of beat
note between the two lasers is shown in Figure 2.

The reference laser is an optical frequency standard based on noise-immune
cavity-enhanced optical heterodyne molecular spectroscopy (NICE-OHMS) [12].
More particularly, the emission frequency of an extended-cavity diode laser (ECDL)
is actively stabilized against the center of a sub-Doppler $H_2^{18}O$ line, under optical
saturation conditions [13]. In this respect, we exploit the fact that NICE-OHMS
provides a dispersion signal without dithering the optical cavity, likely to be employed as an error signal. We should mention that the presence of such a dither was

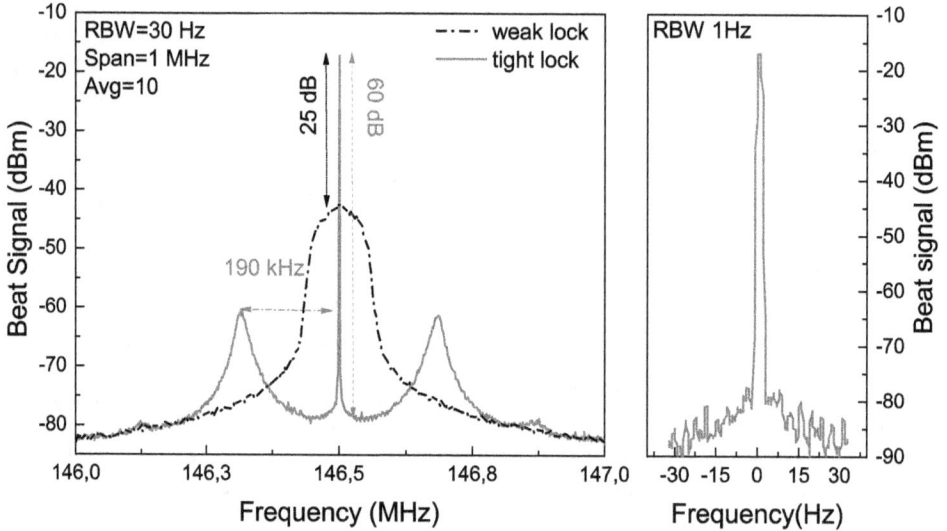

Fig. 2. Examples of beat note between probe and reference lasers under weak and tight lock conditions. The plot on the right side shows the phase-locked beat note with higher resolution, characterized by a signal-to-noise ratio of 60 dB.

a limiting factor in the previous experiment, as widely explained elsewhere [9]. The emission linewidth of the reference laser was carefully determined from the measurement of frequency-noise power spectral density. It turns out that the linewidth of the optical frequency standard amounts to about 7 kHz (full width at half maximum) for an observation time of 1 ms [14].

Laser-gas interaction takes place inside an isothermal cell, actively stabilized at the temperature of the triple point of water. It consists in a spherical, Herriott-type, multiple reflection cell with a maximum path-length of 12 m in a volume of about 400 cm^3. The use of a long-path technique allows us to compensate for the smaller line intensity factor, as compared to the water experiment. The cell was entirely done in stainless steel, with electro-polished inner and outer cavity surfaces. Temperature measurements were taken at the two opposite ends of the cylindrical spacer, by using a pair of capsule-type standard platinum resistance thermometers (SPRTs), placed in the front and in the back of the cell, at 180° between each other. SPRTs were calibrated at INRIM (the National Institute for Research in Metrology) in the temperature interval between the triple point of mercury and the indium freezing point. Temperature stability (as determined over one full day at the temperature of the triple point of water) was found to be at the level of 0.1 mK, while temperature homogenity was better than 1 mK. These extraordinary performances were ensured by a sophisticated system, essentially made of two cylindrical chambers, one inside the other, the multipass cell being housed inside the inner one. In order to obtain acoustic and thermal insulation from the outside environment, the two chambers are kept under vacuum conditions. Two independent temperature stabilization stages were used: an auxiliary thermostat and a fine heating control, the former acting on the inner chamber (exploiting the circulation of a thermal fluid), the latter on the multipass cell (through a constantan heater wire, uniformly wrapped around the cell body).

As for the intensity stabilization of the probe laser, we adopted the same configuration of past experiments [7, 8, 15]. Two identical ultralow-noise preamplified InGaAs photodiodes were used, the former to produce the reference signal that is required for the intensity stabilization, the latter to measure the power on the output of the isothermal cell.

Highly linear, accurate and reproducible frequency scans of the probe ECDL around a given center frequency were effectively done by continuously tuning the offset rf frequency. In the adopted configuration, these scans consisted of 3520 steps of 1 MHz each, with a step-by-step acquisition time of 100 ms.

The pressure of the 99.6% pure C_2H_2 sample was measured by means of a 100-Torr capacitance manometer, with an accuracy of 0.25% of the reading. A turbomolecular pump was used to periodically evacuate the isothermal cell and create high-purity conditions.

3. Results

Figure 3 shows example spectra that were recorded at different gas pressures, with a cell's temperature at the triple point of water. In a first measurement series, twenty repeated acquisitions were performed for each of the four pressures. The small decrease of the transmitted power at the low-frequency edge of the scan is due to a neighboring line, which is not reported in the HITRAN database [16]. It can be ascribed to the $^{13}CH^{12}CH$ molecule, by looking into the line list of Ref. 17.

One of the major efforts of this work is line fitting to retrieve the Doppler width with the highest accuracy. Differently from previous works, in which individual fits were performed, a global analysis approach has been applied to simultaneously fit a manifold of experimental profiles across a given range of pressures, sharing a restricted number of unknown parameters, including the Doppler width. The nonlinear least-squares fitting procedure was implemented under the MATLAB environment using the Levenberg-Marquardt optimization algorithm. As explained elsewhere [18], the global analysis is expected to reduce the uncertainty associated to the line shape model. This approach should also reduce fluctuations resulting from statistical correlations among free parameters, as clearly demonstrated in Ref. 19.

Another key point of any DBT experiment is the choice of the line shape model. We adopted the partially-correlated quadratic speed-dependent hard collision model, in which speed dependence is considered in a quadratic form [20, 21]. Very recently, this model has been proposed as a sort of universal profile to be adopted for high-resolution spectroscopy in the gas phase [22]. For simplicity, it was called Hartmann-Tran profile (HTP). The HTP model was judged to be sophisticated enough to capture the various collisional perturbations to the isolated line

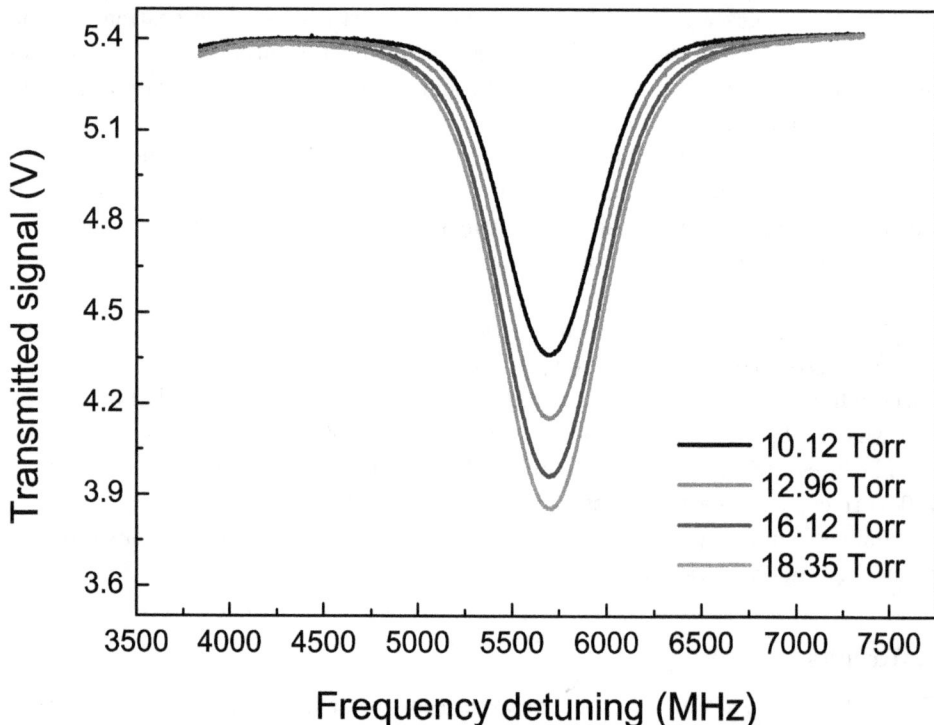

Fig. 3.   Examples of C_2H_2 transmission spectra.

shape, including narrowing effects that result from molecular confinement (namely, the averaging effect of velocity-changing collisions) and speed-dependence of collisional broadening. In addition, this model takes into account the partial correlation between velocity-changing collisions and dephasing collisions (leading to line broadening). Such correlation is described by a parameter η that ranges between 0 and 1.

So far, the HTP profile has not been tested for the aims of DBT experiments. It is worth noting that in the quadratic approximation the collisional width is written as [23]

$$\Gamma^Q(v) = \Gamma_0 \left\{ 1 + a_w \left[\left(\frac{v}{v_0} \right)^2 - \frac{3}{2} \right] \right\},\tag{2}$$

where Γ_0 is the average collisional width, v_0 is the most probable speed of the absorbing molecules, and a_w is a characteristic factor of the speed dependence.

Preliminary results are extremely encouraging. In Figure 4, we report the residuals from the global fitting of the four spectra of Figure 3. The adopted model is capable of reproducing the experimental spectra within the noise, the root-mean-squares value of the residuals being equal to about 1.5 mV, at any pressure. The parameter vector consists of elements of two different types: Those shared among different spectra and those that are characteristic of individual spectra. The Doppler

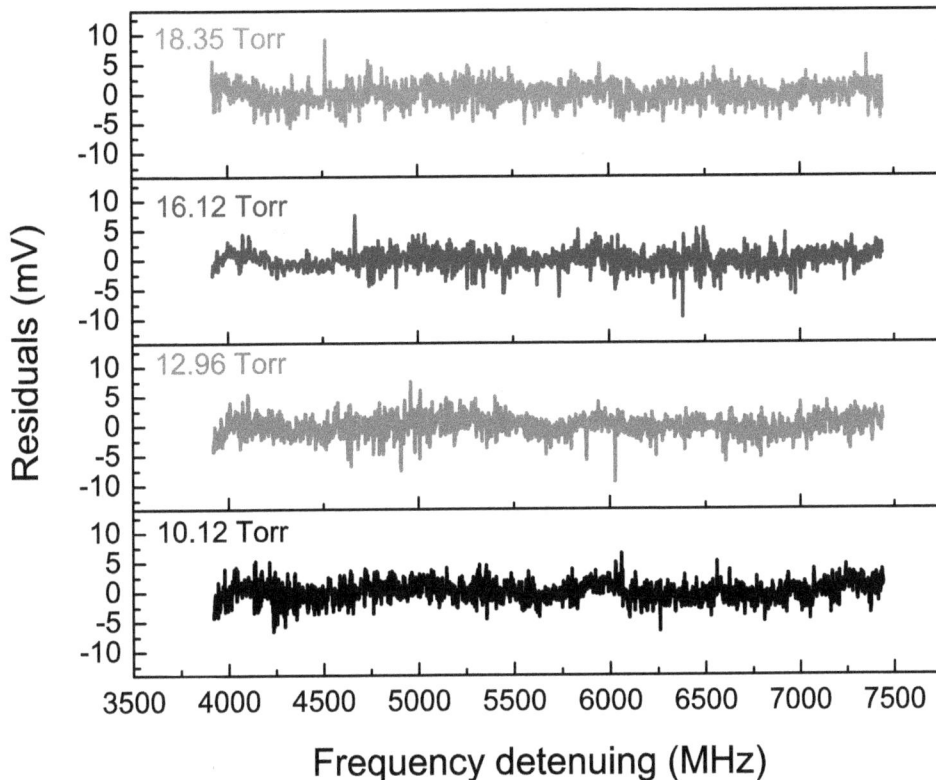

Fig. 4. Fits residuals resulting from the application of the global approach using the HTP profile.

width, the line-center frequency, the velocity-changing collision frequency per unit pressure, the pressure broadening and shifting coefficients, and the a_w parameter are of the first category, while the integrated absorbance and two baseline parameters (accounting for a possible residual variation of the probe laser power) belong to the second one. Therefore, the number of parameters was 18. The global fitting procedure was applied to 20 datasets, each of them consisting of four spectra. The spectroscopic temperatures (as retrieved from the Doppler widths, using the CODATA recommended value for the Boltzmann constant [24]) are shown in Figure 5. The weighted mean of these values gives 273.13 (3) K, in good agreement with the set-point, namely, 273.1602 (3) K.

4. Conclusions

In the third generation experiment at the Second University of Naples, the Boltzmann constant will be retrieved from the shape of a vibration-rotation line of acetylene at 1.39 μm. This molecular species already attracted the interest of the international community of fundamental metrology, when looking for a primary wavelength standard for the important field of optical telecommunications [25].

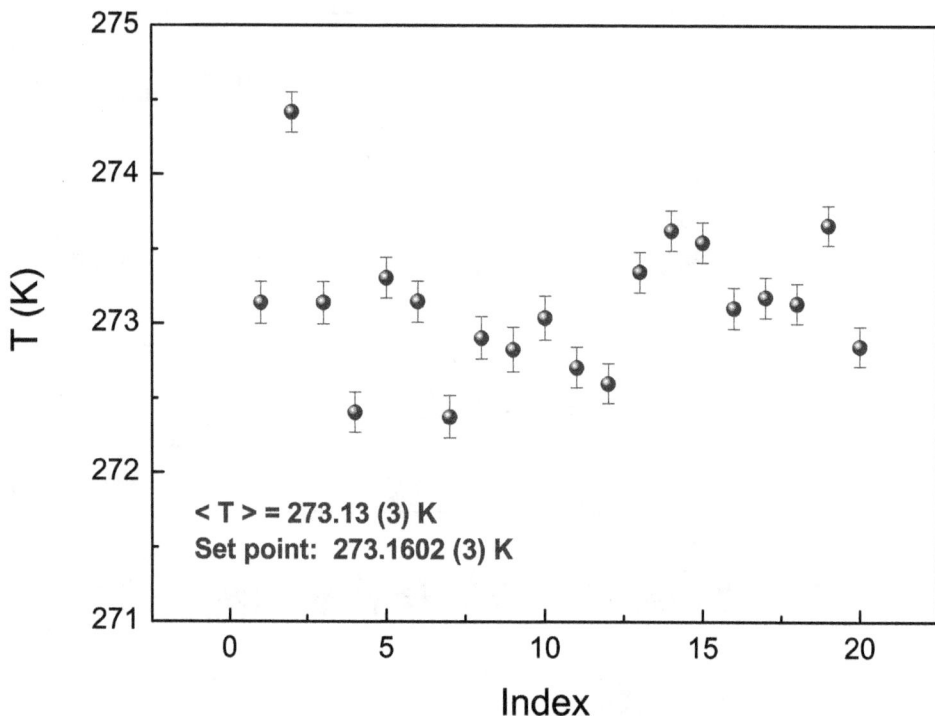

Fig. 5. Spectroscopic determinations of the gas temperature. Each point results from the global fitting of four spectra recorded at different presures.

A first test over 80 spectra allowed us to retrieve the gas temperature with a precision of 110 ppm, in good agreement with the expected value. The test was repeated by using eleven different values of the gas pressure, rather than four. Also in this case, the global fitting procedure was applied to 20 datasets, with the only difference that each of them consisted of eleven spectra. The resulting statistical uncertainty was reduced down to 40 ppm. This is already better than the recent result obtained by Hashemi *et al.*, who probed the C_2H_2 P(25) transition of the $\nu_1 + \nu_3$ band at 1.54 μm [26].

A further increase of the number of spectra would surely lead to a better precision. Nevertheless, an improvement of the signal-to-noise ratio appears to be an indispensable prerequisite, in order to avoid managing thousands of spectra. In this respect, the present value of roughly 3500 should be improved by at least a factor of 3 to reach the same level of the water experiment [7]. To this end, the use of a nonpolar molecule is surely advantageous. In fact, collisions are less effective, while molecule adsorption and desorption from the cell walls are strongly reduced. This feature makes it possible to perform spectral averaging over long times and increase the signal-to-noise ratio without the risk of line distortion that may be caused by a pressure variation arising from the interaction with the cell.

As for systematical deviations, the significant improvement in the spectral purity of the two lasers allows us to neglect the perturbation of the line profile associated to the so-called instrumental width. Finally, the global fitting approach appears to be effective, reliable and robust. Most importantly, it does not require any preliminary knowledge about collisional parameters, circumstance that should lead to a reduction of the uncertainty associated to the line shape model. This latter feature, however, requires further (and deeper) investigations, also testing other profiles.

Acknowledgements

The authors are grateful to Andrea Merlone for the precious contribution regarding temperature traceability of the isothermal cell. The authors gratefully acknowledge funding from EURAMET through the EMRP projects No. SIB01-REG3 and SIB01-REG4, within the Ink (Implementing the new Kelvin) Project coordinated by Graham Machin. The EMRP is jointly funded by the EMRP participating countries within EURAMET and the European Union.

References

[1] M. R. Moldover, J. P. M. Trusler, T. J. Edwards, J. B. Mehl and R. S. Davis, Measurement of the universal gas constant R using a spherical acoustic resonator, *Phys. Rev. Lett.* **60**, 249 (Jan 1988).

[2] M. R. Moldover, R. M. Gavioso, J. B. Mehl, L. Pitre, M. de Podesta and J. T. Zhang, Acoustic gas thermometry, *Metrologia* **51**, p. R1 (2014).

[3] M. de Podesta, R. Underwood, G. Sutton, P. Morantz, P. Harris, D. F. Mark, F. M. Stuart, G. Vargha and G. Machin, A low-uncertainty measurement of the Boltzmann constant, *Metrologia* **50**, p. 354 (2013).

[4] C. Gaiser, T. Zandt, B. Fellmuth, J. Fischer, O. Jusko and W. Sabuga, Improved determination of the Boltzmann constant by dielectric-constant gas thermometry, *Metrologia* **50**, p. L7 (2013).

[5] C. J. Bordé, Atomic clocks and inertial sensors, *Metrologia* **39**, p. 435 (2002).

[6] C. J. Bordé, Base units of the si, fundamental constants and modern quantum physics, *Phyl. Trans. R. Soc. A* **363**, p. 2177 (2005).

[7] L. Moretti, A. Castrillo, E. Fasci, M. D. De Vizia, G. Casa, G. Galzerano, A. Merlone, P. Laporta and L. Gianfrani, Determination of the Boltzmann constant by means of precision measurements of $H_2^{18}O$ line shapes at 1.39 μm, *Phys. Rev. Lett.* **111**, p. 060803 (Aug 2013).

[8] G. Casa, A. Castrillo, G. Galzerano, R. Wehr, A. Merlone, D. Di Serafino, P. Laporta and L. Gianfrani, Primary gas thermometry by means of laser−absorption spectroscopy: Determination of the Boltzmann constant, *Phys. Rev. Lett.* **100**, p. 200801 (May 2008).

[9] A. Castrillo, L. Moretti, E. Fasci, M. D. Vizia, G. Casa and L. Gianfrani, The Boltzmann constant from the shape of a molecular spectral line, *Journal of*

Molecular Spectroscopy **300**, 131 (2014), Spectroscopic Tests of Fundamental Physics.

[10] E. Fasci, M. D. D. Vizia, A. Merlone, L. Moretti, A. Castrillo and L. Gianfrani, The Boltzmann constant from the $H_2^{18}O$ vibration−rotation spectrum: Complementary tests and revised uncertainty budget, *Metrologia* **52**, p. S233 (2015).

[11] D. Jacquemart, N. Lacome, J.-Y. Mandin, V. Dana, H. Tran, F. Gueye, O. Lyulin, V. Perevalov and L. Rgalia-Jarlot, The IR spectrum of $^{12}C_2H_2$: Line intensity measurements in the 1.4 μm region and update of the databases, *Journal of Quantitative Spectroscopy and Radiative Transfer* **110**, 717 (2009).

[12] J. Ye, L.-S. Ma and J. L. Hall, Sub-doppler optical frequency reference at 1.064 μm by means of ultrasensitive cavity-enhanced frequency modulation spectroscopy of a C_2HD overtone transition, *Opt. Lett.* **21**, 1000 (Jul 1996).

[13] H. Dinesan, E. Fasci, A. Castrillo and L. Gianfrani, Absolute frequency stabilization of an extended-cavity diode laser by means of noise-immune cavity-enhanced optical heterodyne molecular spectroscopy, *Opt. Lett.* **39**, 2198 (Apr 2014).

[14] H. Dinesan, E. Fasci, A. D'Addio, A. Castrillo and L. Gianfrani, Characterization of the frequency stability of an optical frequency standard at 1.39 μm based upon noise-immune cavity-enhanced optical heterodyne molecular spectroscopy, *Opt. Express* **23**, 1757 (Jan 2015).

[15] G. Casa, D. A. Parretta, A. Castrillo, R. Wehr and L. Gianfrani, Highly accurate determinations of CO_2 line strengths using intensity-stabilized diode laser absorption spectrometry, *The Journal of Chemical Physics* **127**, p. 084311 (2007).

[16] L. Rothman, I. Gordon, Y. Babikov, A. Barbe, D. C. Benner, P. Bernath, M. Birk, L. Bizzocchi, V. Boudon, L. Brown, A. Campargue, K. Chance, E. Cohen, L. Coudert, V. Devi, B. Drouin, A. Fayt, J.-M. Flaud, R. Gamache, J. Harrison, J.-M. Hartmann, C. Hill, J. Hodges, D. Jacquemart, A. Jolly, J. Lamouroux, R. L. Roy, G. Li, D. Long, O. Lyulin, C. Mackie, S. Massie, S. Mikhailenko, H. Mller, O. Naumenko, A. Nikitin, J. Orphal, V. Perevalov, A. Perrin, E. Polovtseva, C. Richard, M. Smith, E. Starikova, K. Sung, S. Tashkun, J. Tennyson, G. Toon, V. Tyuterev and G. Wagner, The HITRAN2012 molecular spectroscopic database, *Journal of Quantitative Spectroscopy and Radiative Transfer* **130**, 4 (2013).

[17] S. Robert, B. Amyay, A. Fayt, G. D. Lonardo, L. Fusina, F. Tamassia and M. Herman, Vibration-rotation energy pattern in acetylene: $^{13}CH^{12}CH$ up to 10120 cm^{-1}, *The Journal of Physical Chemistry A* **113**, 13251 (2009), PMID: 19921941.

[18] P. Amodio, M. D. De Vizia, L. Moretti and L. Gianfrani, Investigating the ultimate accuracy of doppler-broadening thermometry by means of a global fitting procedure, *Phys. Rev. A* **92**, p. 032506 (Sep 2015).

[19] P. Amodio, L. Moretti, A. Castrillo and L. Gianfrani, Line-narrowing effects in the near-infrared spectrum of water and precision determination of spectroscopic parameters, *The Journal of Chemical Physics* **140**, p. 044310 (2014).

[20] N. Ngo, D. Lisak, H. Tran and J.-M. Hartmann, An isolated line-shape model to go beyond the voigt profile in spectroscopic databases and radiative transfer codes, *Journal of Quantitative Spectroscopy and Radiative Transfer* **129**, 89 (2013).

[21] H. Tran, N. Ngo and J.-M. Hartmann, Efficient computation of some speed-dependent isolated line profiles, *Journal of Quantitative Spectroscopy and Radiative Transfer* **129**, 199 (2013).

[22] J. Tennyson, P. F. Bernath, A. Campargue, A. G. Császár, L. Daumont, R. R. Gamache, J. T. Hodges, D. Lisak, O. V. Naumenko, L. S. Rothman, H. Tran, N. F. Zobov, J. Buldyreva, C. D. Boone, M. D. D. Vizia, L. Gianfrani, J.-M. Hartmann, R. McPheat, D. Weidmann, J. Murray, N. H. Ngo and O. L. Polyansky, Recommended isolated-line profile for representing high-resolution spectroscopic transitions (IUPAC technical report), *Pure and Applied Chemistry* **86**, 1931 (2014).

[23] L. Gianfrani, Highly-accurate line shape studies in the near-IR spectrum of $H_2{}^{18}O$: Implications for the spectroscopic determination of the Boltzmann constant, *Journal of Physics: Conference Series* **397**, p. 012029 (2012).

[24] P. J. Mohr, B. N. Taylor and D. B. Newell, CODATA recommended values of the fundamental physical constants: 2010, *Rev. Mod. Phys.* **84**, 1527 (Nov 2012).

[25] T. J. Quinn, Practical realization of the definition of the metre, including recommended radiations of other optical frequency standards (2001), *Metrologia* **40**, p. 103 (2003).

[26] R. Hashemi, C. Povey, M. Derksen, H. Naseri, J. Garber and A. Predoi-Cross, Doppler broadening thermometry of acetylene and accurate measurement of the Boltzmann constant, *The Journal of Chemical Physics* **141**, p. 214201 (2014).

Antiferromagnetism with Ultracold Atoms

Randall G. Hulet*, Pedro M. Duarte, Russell A. Hart and Tsung-Lin Yang

Department of Physics and Astronomy, Rice University,
Houston, Texas 77005, USA
* E-mail: randy@rice.edu
atomcool.rice.edu

We use ultracold spin–1/2 atomic fermions (^{6}Li) to realize the Hubbard model on a three-dimensional (3D) optical lattice. At relatively high temperatures and at densities near half-filling, we show that the gas forms a Mott insulator with unordered spins. To observe antiferromagnetic order that is predicted to occur at lower temperatures, we developed the compensated optical lattice method to evaporatively cool atoms in the lattice. This cooling has enabled the detection of short-range magnetic order by spin-sensitive Bragg scattering of light.

Keywords: Hubbard model; quantum magnetism; fermions; optical lattice.

1. Introduction

Cold atomic gases have emerged as a versatile new platform for applications to fundamental many-body physics [1]. This versatility arises from a remarkable ability to precisely control nearly every relevant system parameter. Most notably, the density, interaction, the presence or complete absence of disorder, and dimensionality can all be controlled. The one glaring, and severely limiting exception to this list, is temperature. While the absolute temperatures of cold gases are lower than for any other man-made or naturally occurring physical system, the absolute temperature is not the most important quantity. Rather, the relevant energy scale is set by the Fermi temperature T_F, which determines the absolute temperature T where phase transitions to superfluid or quantum magnetic states occur. (For bosons, the analogous energy scale is the Bose-Einstein condensation temperature, which is nearly identical to T_F). The Fermi energy in cold gases is constrained to be about $1\,\mu$K by the need to keep densities low enough to minimize molecular recombination. In contrast, the Fermi temperature of a high-temperature superconductor, at $\sim$10,000 K, is nearly 10^{10} times higher. In these materials, a transition temperature as high as 100 K is only 1% of T_F, well below the relative temperatures achieved in state-of-the-art atomic fermion experiments, for which the most advanced cooling methods, performed by evaporative cooling of atoms in traps, have produced $T/T_F \simeq 0.04$.

For cold atomic gases to fully attain their potential to realize and characterize new quantum states of matter requires that we develop methods for cooling to ever lower temperatures. In our work, we are particularly motivated to improve our understanding of high-temperature superconductors, where the ultimate goal is to create new materials with even higher transition temperatures than currently possible.

Our specific interest, and an ideal test-bed for our cooling methods, is the Hubbard model, which was originally proposed to describe the magnetic and conduction properties of electrons in transition metal oxides. Shortly after the discovery of high-temperature superconductors, Anderson suggested that the Hubbard model may also contain the essential ingredients to describe their remarkable properties [2]. The Hubbard model has now become one of the most important models of strongly correlated matter, and was recognized early-on as a target for emulation with ultracold atoms on an optical lattice [3, 4]. The model itself is deceptively simple, describing a spin-1/2 Fermi gas on a lattice by its hopping rate t, and an on-site interaction energy U. Only the ground band need be considered when both T and U are much less than the band gap. In this case, the Pauli principle forbids atoms of the same spin from occupying the same lattice site.

Despite its simplicity, the solutions of the Hubbard Hamiltonian are remarkably complex depending on the sign and magnitude of the ratio U/t, as well as the density n. Similarly, the phase diagram of a typical high-T_c material exhibits a variety of complex behaviors, including antiferromagnetism, a poorly understood pseudo-gap region in which there are pair correlations, but no long-range order, and at very low temperatures, d-wave superconductivity. It is unknown whether the Hubbard model is sufficient to provide a unifying theory of high-T_c materials, because it cannot be solved with analytic or numerical methods since the basis size, which scales as 2^N, where N is the total number of particles, quickly overwhelms the computational capability of any current or envisioned digital computer. By using cold atoms on a lattice, we create an analog quantum computer specifically designed to solve the Hubbard model.

2. Experiment

The experiment utilizes the $|F = 1/2; m_F = +1/2\rangle$ and $|F = 1/2; m_F = -1/2\rangle$ hyperfine states of ^{6}Li, which we label $|\uparrow\rangle$ and $|\downarrow\rangle$, respectively. Our methods have been described previously [5, 6]. We use all-optical methods [7] to confine and cool the atoms, before evaporatively cooling them in a crossed-beam optical trap. These atoms are then loaded into a simple cubic optical lattice. The temperature of the atoms in the trap, prior to loading them into the lattice, is measured to be $T/T_F = 0.04 \pm 0.02$ by fitting the density distribution after time of flight.

The optical lattice is formed by three retroreflected red-detuned ($\lambda = 1064$ nm) Gaussian laser beams of depth $V_0 = 7 E_r$, where $E_r = \frac{\hbar^2 \pi^2}{2ma^2} = 1.4 \, \mu$K is the recoil energy, m is the mass of the atoms, and $a = \lambda/2$ is the lattice spacing. We use the broad Feshbach resonance in ^{6}Li at 832 G [8, 9] to set the on-site interaction strength, U. Typical experimental parameters are $V_0 = 7 E_r$, $t = 0.038 E_r$, and $U = 0.38 E_r$, corresponding to an s-wave scattering length of $260 \, a_0$.

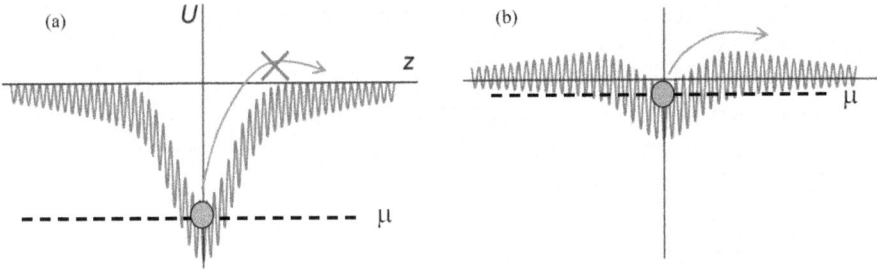

Fig. 1. (a) Optical lattice with the usual Gaussian confinement, which inhibits evaporation; (b) Compensated optical lattice showing evaporation.

2.1. Compensated Optical Lattice

Temperatures of fermions in traps can usually be cooled to a factor of ~3-4 below that in optical lattices. There are several factors that contribute to this situation, including technical noise sources in lattices that are more pernicious than in traps, but the primary reason is that the depth of a trap may be adjusted such that the chemical potential, μ, lies just below the lip of the trap. A small, but steady rate of evaporation mitigates heating to produce relative temperatures of $T/T_F \simeq 0.04$. In a standard lattice, however, the Gaussian beams create a confining envelope that cannot be adjusted without also affecting the lattice depth, V_0. Since μ lies far below the lip of the confinement potential at densities of $n = 1$ (one atom per lattice site or "half-filling"), evaporation is impeded by the small Boltzmann factor,

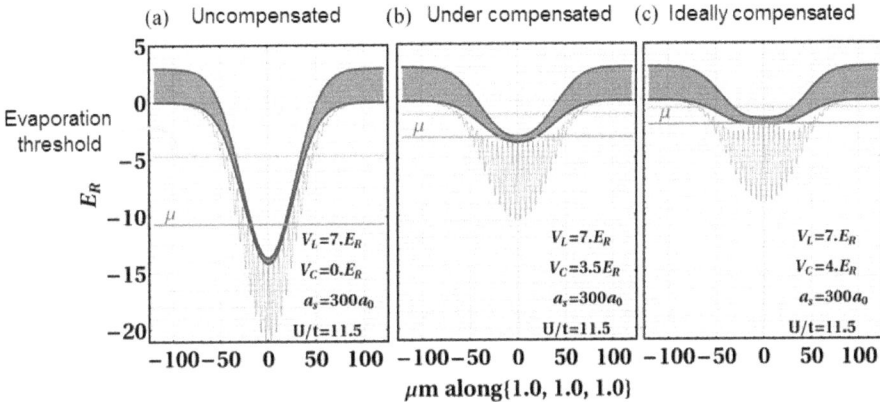

Fig. 2. (a) Uncompensated optical lattice; (b) Under compensated lattice; and (c) Ideally compensated lattice. The oscillating lines represents the lattice, where the lattice constant is exaggerated for clarity. The dark region is the lowest Bloch band. Compensation can both bring μ close to the evaporation threshold, and with the proper relative beam sizes, it can also flatten the band, as shown in (c). In all three panels μ corresponds to a central density of $n = 1$.

as shown in Fig. 1(a). We introduce additional degrees of freedom, provided by blue-detuned (532 nm) anti-confining compensation beams, to control the depth of confinement independently of V_0. These compensation beams overlap each of the lattice beams but are not themselves retroreflected [6, 10]. This provides the ability to tune the depth of confinement such that atoms may evaporate, as shown in Fig. 1(b). With the additional freedom to choose the Gaussian waists of the compensating beams relative to the lattice beams, it is also possible to flatten the lowest band, and thereby increase the volume fraction where μ is nearly constant [10]. The effect of compensation on the band structure is illustrated in Fig. 2.

3. Results

3.1. *Mott Insulator*

The Hubbard model for densities near $n = 1$ and for $T \ll U$ describes a system which undergoes a smooth crossover to a Mott insulating regime, characterized by a suppression of density fluctuations and a reduction of the number of doubly occupied sites. The suppression of density fluctuations implies a reduction of the compressibility [11]. Figure 3 shows the resulting plateaus in the measured trapped density distributions at $n = 1$ when U/T becomes sufficiently large. We have used these density distributions to extract the local compressibility of the trapped gas as a function of n and have shown that this quantity can provide accurate thermometry in the lattice down to temperatures $T \simeq t$. The Mott transition in the Fermi-Hubbard model has been studied in previous experiments by measuring the variation of the bulk double occupancy with atom number [12] and the response of the cloud radius to changes in external confinement [13], both of which are related to the global compressibility, and are severely suppressed for large interactions.

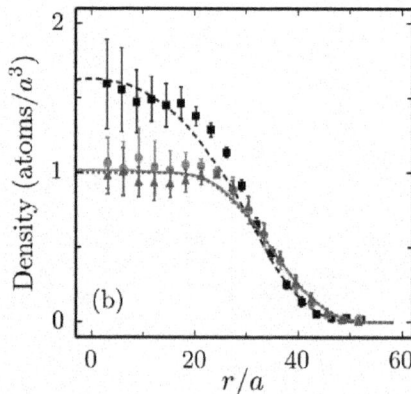

Fig. 3. Density distributions showing the development of the Mott plateau at $n = 1$ with increasing U/t. These 3D distributions are obtained from the column density images by an Abel transform. The squares, dots, and triangles correspond to $U/t = 3.1$, 11.1, and 14.5, respectively. (Reprinted from Ref. 5).

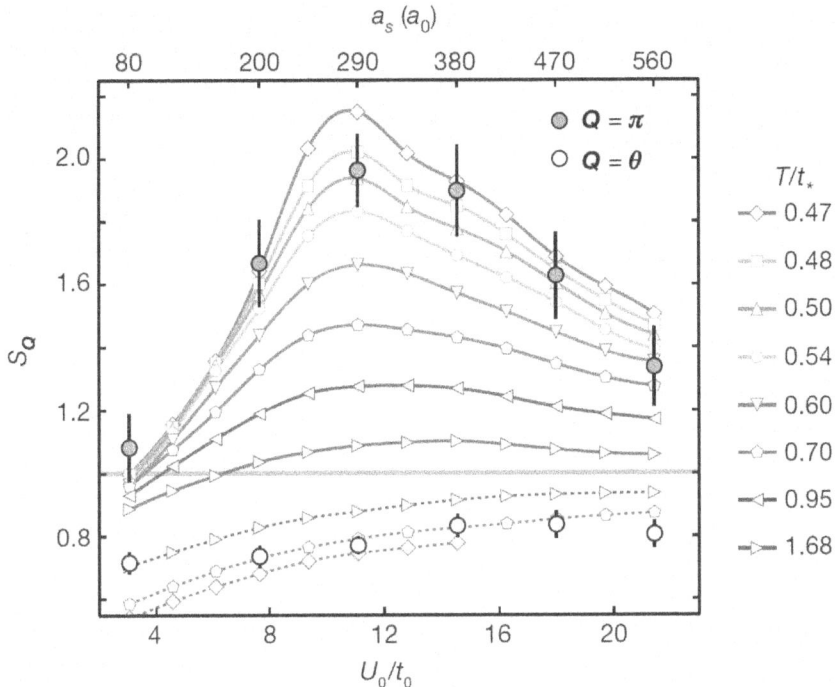

Fig. 4. Bragg scattering signal. Data shown by filled dots corresponds to the spin-structure factor in the coherently scattered direction (satisfying the Bragg condition with a reciprocal lattice vector $Q = (2\pi/a)(\frac{1}{2}, \frac{1}{2}, \frac{1}{2}))$. The different symbols and lines are the results of the QMC calculations for the different temperatures indicated. Data given by open circles correspond to a non-Bragg angle, θ. $S_\theta < 1$ because doubly-occupied sites scatter no light and there is no coherent enhancement in this direction. $S_Q = 1$ in the high-temperature limit for any direction. (Reprinted from Ref. 6).

3.2. *Antiferromagnetic Order*

At very low temperatures, the Mott insulator undergoes a phase transition to an antiferromagnetic (AFM) state. This transition occurs at the Néel temperature, T_N ($\sim 4t^2/U$ for $U \gg t$). T_N is more than a factor of two below the lowest temperatures previously attained in a isotropic 3D lattice [14, 15]. In our experiment, we use Bragg scattering of near-resonant light [16] to detect the onset of AFM order [6]. Detection of magnetic order requires that the light scattering be spin-dependent. We accomplished this by tuning the laser between the transition frequencies of the two states, $|\uparrow\rangle$ and $|\downarrow\rangle$, of our spin-1/2 system.

The strength of the Bragg signal, which is directly proportional to the spin structure factor, S_π, is shown as the solid dots in Fig. 4 as a function of U/t [6]. As expected from quantum Monte Carlo (QMC) calculations, S_π exhibits a broad maximum centered near $U/t \approx 10$ [17]. The results of QMC for various temperatures are also shown in Fig. 4. The QMC calculations are performed for a homogeneous density distribution and then averaged over the measured density distribution using the local density approximation. The calculations demonstrate that the Bragg signal is extremely sensitive to temperature in this regime, and thus the combination of Bragg scattering and QMC provides sensitive thermometry in a regime where previously there was none. Comparison of the data with QMC indicates that the temperature is $T/t = 0.51 \pm 0.06$, where the uncertainty is due to the statistical error in the measured S_π. In terms of T_N, the temperature is $T/T_N = 1.42 \pm 0.16$.

4. Conclusions and outlook

In conclusion, we have detected short-range antiferromagnetic correlations in the Hubbard model. At 1.4 T_N, the correlations are still short-range. We are able to observe these correlations because the compensated optical lattice produces temperatures more than a factor of 2 lower than previously attained for fermions in a 3D optical lattice, and because Bragg scattering of light is an extremely sensitive detector. We must cool further in order to get below T_N, and more generally, to open up the study of novel quantum states of matter that have yet to be created. We are optimistic that the compensated lattice method can be pushed further to optimize evaporative cooling in the lattice and to flatten the ground band more completely. To do so, we are using a spatial light modulator to produce compensation intensity distributions that are more complex than just the simple Gaussian beams employed here. In addition, we are exploring ways to use blue-detuned light to imprint metallic regions with $n < 1$ on the Mott plateau as a way of storing entropy or conducting it away from the low-entropy, but insulating Mott phase.

Acknowledgments

We gratefully acknowledge the invaluable contributions by our theory collaborators T. Paiva, E. Khatami, R. T. Scalettar, N. Trivedi, and D. A. Huse. This work was supported by NSF grant No. PHY-1408309, the ONR, the Welch Foundation (Grant No. C-1133), and an ARO-MURI grant No. W911NF-14-1-003.

References

[1] I. Bloch, J. Dalibard and W. Zwerger, Many-body physics with ultracold gases, *Rev. Mod. Phys.* **80**, 885 (2008).

[2] P. W. Anderson, The resonating valence bond state in La_2CuO_4 and superconductivity, *Science* **235**, 1196 (1987).

[3] W. Hofstetter, J. I. Cirac, P. Zoller, E. Demler and M. D. Lukin, High-temperature superfluidity of fermionic atoms in optical lattices, *Phys. Rev. Lett.* **89**, p. 220407 (2002).

[4] D. Jaksch and P. Zoller, The cold atom Hubbard toolbox, *Annals of Physics* **315**, 52 (2005).

[5] P. M. Duarte, R. A. Hart, T.-L. Yang, X. Liu, T. Paiva, E. Khatami, R. T. Scalettar, N. Trivedi and R. G. Hulet, Compressibility of a fermionic Mott insulator of ultracold atoms, *Phys. Rev. Lett.* **114**, p. 070403 (2015).

[6] R. A. Hart, P. M. Duarte, T.-L. Yang, X. Liu, T. Paiva, E. Khatami, R. T. Scalettar, N. Trivedi, D. A. Huse and R. G. Hulet, Observation of antiferromagnetic correlations in the Hubbard model with ultracold atoms, *Nature* **519**, 211 (2015).

[7] P. M. Duarte, R. A. Hart, J. M. Hitchcock, T. A. Corcovilos, T.-L. Yang, A. Reed and R. G. Hulet, All-optical production of a lithium quantum gas using narrow-line laser cooling, *Phys. Rev. A* **84**, p. 061406 (2011).

[8] M. Houbiers, H. T. C. Stoof, W. I. McAlexander and R. G. Hulet, Elastic and inelastic collisions of ^{6}Li atoms in magnetic and optical traps, *Phys. Rev. A* **57**, R1497 (1998).

[9] G. Zürn, T. Lompe, A. N. Wenz, S. Jochim, P. S. Julienne and J. M. Hutson, Precise characterization of ^{6}Li Feshbach resonances using trap-sideband-resolved rf spectroscopy of weakly bound molecules, *Phys. Rev. Lett.* **110**, p. 135301 (2013).

[10] C. J. M. Mathy, D. A. Huse and R. G. Hulet, Enlarging and cooling the Néel state in an optical lattice, *Phys. Rev. A* **86**, p. 023606 (2012).

[11] L. De Leo, C. Kollath, A. Georges, M. Ferrero and O. Parcollet, Trapping and cooling fermionic atoms into Mott and Néel states, *Phys. Rev. Lett.* **101**, p. 210403 (2008).

[12] R. Jördens, N. Strohmaier, K. Günter, H. Moritz and T. Esslinger, A Mott insulator of fermionic atoms in an optical lattice, *Nature* **455**, 204 (2008).

[13] U. Schneider, L. Hackermüller, S. Will, T. Best, I. Bloch, T. A. Costi, R. W. Helmes, D. Rasch and A. Rosch, Metallic and insulating phases of repulsively interacting fermions in a 3D optical lattice, *Science* **322**, 1520 (2008).

[14] D. Greif, T. Uehlinger, G. Jotzu, L. Tarruell and T. Esslinger, Short-range quantum magnetism of ultracold fermions in an optical lattice, *Science* **340**, 1307 (2013).

[15] J. Imriška, M. Iazzi, L. Wang, E. Gull, D. Greif, T. Uehlinger, G. Jotzu, L. Tarruell, T. Esslinger and M. Troyer, Thermodynamics and magnetic properties of the anisotropic 3D Hubbard model, *Phys. Rev. Lett.* **112**, p. 115301 (2014).

[16] T. A. Corcovilos, S. K. Baur, J. M. Hitchcock, E. J. Mueller and R. G. Hulet, Detecting antiferromagnetism of atoms in an optical lattice via optical Bragg scattering, *Phys. Rev. A* **81**, p. 013415 (2010).

[17] T. Paiva, Y. L. Loh, M. Randeria, R. T. Scalettar and N. Trivedi, Fermions in 3D optical lattices: Cooling protocol to obtain antiferromagnetism, *Phys. Rev. Lett.* **107**, p. 086401 (2011).

Optical Clocks with Trapped Ions:
Establishing the 10^{-18} Uncertainty Range

Ekkehard Peik

Physikalisch-Technische Bundesanstalt
Bundesallee 100, 38116 Braunschweig, Germany
E-mail: ekkehard.peik@ptb.de
www.ptb.de/time/

Great progress has been made in the development of atomic clocks at optical frequencies and a number of different systems have recently been evaluated with systematic uncertainties in the range of 10^{-18}. Comparisons of frequencies and frequency ratios are now performed to affirm these evaluations, and can also lead to improved tests of relativity and fundamental principles. This paper gives a brief overview of recent improvements in the generation and transfer of stable optical frequencies and focuses on recent work done with the ^{171}Yb$^+$ frequency standards at PTB.

Keywords: optical clock, laser frequency stabilization, light shift, frequency transfer.

1. Introduction: Vision to reality

The potential of a highly precise optical clock or frequency standard with a relative uncertainty in the range of 10^{-18} was first put forward by Hans Dehmelt in 1981: *"Thus the current promise of an atomic line spectral resolution of 1 part in 10^{18} or 10^8 times better than achieved to date may be realized in the not too far future."* [1]. Envisioning an improvement by about eight orders of magnitude through the use of a single trapped and laser-cooled ion, the idea was regarded as visionary but was also considered as highly optimistic by some experts of laser spectroscopy. Nevertheless it rapidly became established as the long-term goal and challenge for scientists working on optical frequency standards. The level of the challenge can be seen by looking at some simple proportions: One part in 10^{18} corresponds to less than 1 s over the age of the universe. Obtaining such an uncertainty necessitates the counting of optical oscillations at 1×10^{15} Hz over 1000 s without loosing more than a single cycle. In terms of a systematic frequency shift, it corresponds to the relativistic time dilation $v^2/2c^2$ observed for a clock that moves at slow pedestrian speed of $v = 0.4$ m/s.

Dehmelt pointed out that using a transition between levels with vanishing electronic angular momentum presents advantages in avoiding or reducing frequency shifts induced by electric and magnetic fields. He proposed to use the hyperfine-induced $^1S_0 \rightarrow {}^3P_0$ electric dipole transition in Tl$^+$. While this system so far does not seem to have been studied experimentally, the first optical clock that reported an uncertainty evaluation resulting in a value below 10^{-17} in 2010 has been built at NIST using the same type of transition in Al$^+$ [2]. The predominantly scalar interaction of the $^1S_0$ and $^3P_0$ levels with far detuned light fields have also made this type of transition the best choice for the optical lattice clock [3] where atoms are trapped in the interference pattern of a laser that exerts equal light shifts on the upper and lower level of the reference transition. About 40 years after the initial proposals, the field of optical clocks has now seen rapid progress in different systems, with laser-cooled atoms and ions of different elements and with different types of transitions. See Refs. [4, 5] for recent reviews. Until the summer of 2015, five groups have reported systematic uncertainties in the 10^{-18} range. In comparison to primary caesium clocks – caesium fountains with laser-cooled atoms – this represents an improvement in precision by up to two orders of magnitude.

It can be expected that frequency comparisons between these different optical clocks will now allow one to reliably identify possible problems with systematic frequency shifts. The most obvious scientific benefits from such an ensemble of different highly precise clocks will come from improved tests of fundamental physics, like in searches for violations of Einstein's equivalence principle. Clocks based on different physical reference systems can be used for example to test the universality of the gravitational redshift and to search for possible temporal variations of the fundamental coupling constants.

This paper gives a brief overview of recent improvements in the generation and transfer of stable optical frequencies and focuses on recent work done with the ^{171}Yb$^+$ frequency standards at PTB.

2. Frequency-stable lasers

The linewidth and noise power spectrum of the laser that serves as the interrogation oscillator has significant influence on the performance of the clock, especially on the obtainable short-term stability. In the case of interrogating a single ion, quantum projection noise usually dominates the signal-to-noise that is obtainable in a single interrogation. The clock stability can be improved by obtaining a narrower resonance linewidth. In a number of important cases (the $^1S_0 \rightarrow {}^3P_0$ transition in Al$^+$ and especially the Yb$^+$ electric octupole transition, to be discussed below) the natural linewidth due to spontaneous decay of the excited state is smaller than the laser linewidth that has been obtained so far. In these cases the clock is usually operated with an interrogation time that is adapted to the laser coherence, so that improved laser stability may immediately improve the clock. In the case of a lattice clock with many atoms, a more frequency-stable laser can be used to obtain quantum noise-limited stability with a higher atom number.

Most of the stable lasers in use today are based on passive, high-finesse, environmentally-isolated Fabry-Perot resonators. Their frequency stability is limited by mechanical thermal noise in their constituents, with a big contribution coming from the mirror coatings [6]. This noise can be reduced by operation at cryogenic temperature, the use of (crystalline) materials with high mechanical Q-factor, by increasing the mode size on the mirrors for improved averaging of the thermal fluctuations, or by using a longer cavity and thereby reducing the relative contribution from the coating material. Excellent results have recently been obtained with a cavity spacer and the mirror substrates made from single-crystal silicon and operated at 124 K where the silicon thermal expansion coefficient is zero and the silicon mechanical loss is small. With this resonator it was possible to demonstrate a fractional frequency stability below 1×10^{-16} for averaging times between 0.1 and 1 s, supporting a laser linewidth below 40 mHz at 1.5 μm wavelength [7]. A noise level below 1×10^{-16} extending even to averaging times of 1000 s has been obtained with a 48 cm long resonator made from ultra-low expansion glass with optically contacted fused silica mirror substrates [8]. Further progress in the stability is expected when ion beam sputtered dielectric mirror coatings are replaced by compound-semiconductor-based epitaxial crystalline multilayers, which exhibit intrinsically low mechanical loss [9].

While the setup of such a high-performance resonator requires efforts and resources like cryogenics, thermal insulation and vibration control, a single system may be used to provide stability to several reference lasers, even at different wavelengths by using transfer of the stability via a frequency comb. The 1×10^{-16} level of laser stability at 1 s is an important benchmark in the design of an optical clock, because in generating the error signal from the atoms for longer averaging times, the clock instability will follow the characteristic $1/\sqrt{\tau}$ dependence (τ: averaging time) of the atomic quantum noise. Starting from the low 10^{-16} range in 1 s will allow one to reach a statistical uncertainty of 1×10^{-18} within a manageable averaging time of a few hours only.

3. Optical clocks with a single trapped ^{171}Yb$^+$ ion

A partial level scheme of the rare-earth ion Yb$^+$ is shown in Fig. 1. A $^2S_{1/2} \to {}^2P_{1/2}$ transition is used for laser cooling and fluorescence detection and transitions from the ground state to metastable D and F levels serve as references for optical clocks. Recent work has shown several important, partly unexpected advantages of this ion and has made it possible to obtain the lowest systematic uncertainty among the single-ion optical clocks at present.

The isotope ^{171}Yb$^+$ with nuclear spin $I = 1/2$ is used, so that a magnetic field insensitive, non-degenerate $F = 0$ hyperfine level of the ground state is available that can easily be prepared by hyperfine pumping. The relatively high atomic mass leads to smaller Doppler shift at a given temperature. In experiments with trapped Yb$^+$ ions very long storage times exceeding several months can be observed [10], facilitating the long-term continuous operation of the standard. While in other ions chemical reactions with background gas seem to ultimately limit the storage time, this loss process is prevented for Yb$^+$ by the photodissociation of the product ion YbH$^+$ with the 370 nm cooling laser light. Two optical reference transitions in ^{171}Yb$^+$ are being studied at PTB in Germany, NPL in the UK and in

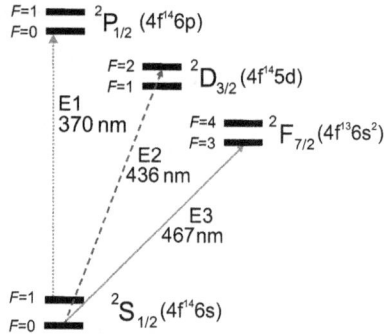

Fig. 1. Partial level scheme of the ^{171}Yb$^+$ ion. The strong $^2S_{1/2} \rightarrow {}^2P_{1/2}$ electric dipole (E1) transition at 370 nm is used to laser-cool the ion. The electric quadrupole (E2) and electric octupole (E3) transition serve as references for optical clocks.

other laboratories: the $^2S_{1/2} - {}^2D_{3/2}$ electric quadrupole (E2) transition [11] and the $^2S_{1/2} - {}^2F_{7/2}$ electric octupole (E3) transition [12]. The octupole transition between the $^2S_{1/2}$ ground state and the lowest excited $^2F_{7/2}$ state is unusal because of its extremely small natural linewidth in the nanohertz range. While allowing for very high resolution, at the limit imposed by noise of the interrogation laser, an associated disadvantage is a significant light shift of the transition frequency [13]. This shift is proportional to the laser intensity so that a π-pulse with Fourier-limited spectral width Δf causes a shift proportional to $(\Delta f)^2$. The shift contains both scalar and tensorial contributions and scales like $0.65(3)$ Hz$^{-1}(\Delta f)^2$ if the polarization and magnetic field orientation are chosen to maximize the excitation probability [14]. Apart from the light shift, the sensitivities of the Yb$^+$ octupole transition frequency to static electric field induced shifts are significantly lower than those of the quadrupole transitions in the alkali-like ions, as has partly been pointed out in theoretical estimates [15] and measured in the frequency standards [14]. Qualitatively, this can be explained by the electronic configuration $(4f^{13}6s^2)$ of the $^2F_{7/2}$ level that consists of a hole in the $4f$ shell that is surrounded by the filled $6s$ shell. The polarizability of the $^2F_{7/2}$ level is close to that of the ground state and smaller than that of the $^2D_{3/2}$ level. Comparing the sensitivity factors for field-induced shifts of the E2 and E3 transitions, the quadratic Zeeman, quadratic Stark (scalar and tensor) and electric quadrupole shifts of the latter are smaller by factors of 26, 8, 62, and 51, respectively, for identical external fields. It is therefore very efficient to use the E2 transition in the same ion as an *in situ* diagnosis for perturbations through external fields and to deduce the much smaller corrections and uncertainties of the E3 frequency from these measurements.

In our first evaluation of a high-accuracy optical clock based on the E3 transition [14], the relative uncertainty of 7.1×10^{-17} was dominated by the contributions from imperfections in a real-time extrapolation of the light shift from the interrogation laser and by the uncertainty in the light shift due to thermal radiation from the

trap and surroundings of the ion. The latter contained contributions from the
uncertainty of the differential polarizability of the ground and excited state and
from limited knowledge of the temperatures of parts of the trap assembly. Solutions
of these three problems have allowed us to reduce the systematic uncertainty into
the low 10^{-18} range.

To treat the problem of the interrogation-related light shift, we use the socalled
Hyper Ramsey spectroscopy [16], an optical excitation scheme that is a generalization of Ramsey's method of separated oscillatory fields and consists of a sequence
of three excitation pulses (see Fig. 2). The pulse sequence is tailored to produce a
resonance signal which is immune to the light shift and other shifts of the transition frequency that are correlated with the interaction with the probe field. With
the excellent control of laser intensity and pulse areas that is possible in a trapped
ion experiment, the scheme is ideally suited to an optical clock based on the E3
transition of ^{171}Yb$^+$.

An excitation spectrum obtained with Ramsey excitation may already show
indications of the presence of light shift: the position and shape of the envelope
reflects the excitation spectrum resulting from one of the pulses, whereas the Ramsey fringes result from coherent excitation with both pulses and the intermediate
dark period. The fringes are less shifted than the envelope, because their shift
is determined by the time average of the intensity. This results in a shifted and
asymmetric Ramsey pattern (see Fig. 2b). This intuitive picture suggests that
the effect of the light shift Δ_L on the spectrum can be compensated by introduc-
ing a frequency step of the probe light $\Delta_S = \Delta_L$ during the interrogation pulses

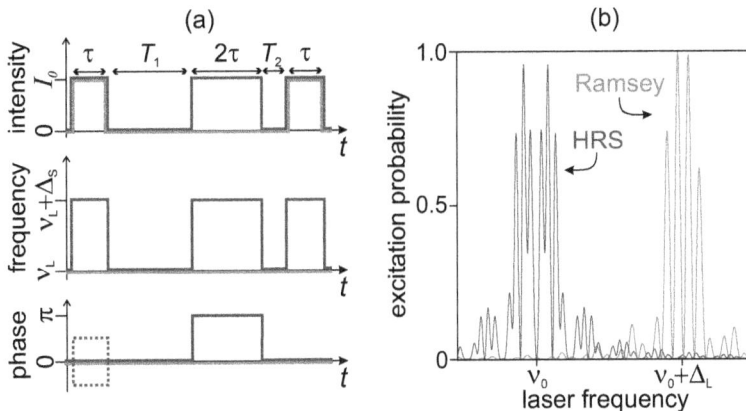

Fig. 2. Pulse sequence (a) and resulting excitation spectrum (b) of the Ramsey and the Hyper-
Ramsey spectroscopy (HRS) excitation scheme. Here ν_L is the probe laser frequency and ν_0 the
unperturbed transition frequency. The laser step frequency Δ_S is assumed to be equal to the
light shift Δ_L and the intensity I_0 is chosen to obtain a pulse area $\pi/2$ for a pulse duration τ. A
discriminator signal can be generated by alternately stepping the phase of the first pulse by $\pm\pi/2$
as indicated by the dotted lines. The spectra are calculated for the parameters $T_1 = 2\tau$, $T_2 = 0$,
$\Delta_L = 4.1/\tau$ with equal dark period durations in both schemes.

whereas the unshifted probe frequency ν_L is tuned across the resonance with the unperturbed atomic frequency ν_0. The scheme can be made insensitive against small changes of the laser intensity or errors in Δ_S by inserting an additional pulse with identical intensity and frequency and with a doubled duration between the Ramsey pulses. Shifting the phase of the additional pulse by π relative to the Ramsey pulses improves the robustness against variations of the pulse area. In contrast to extrapolation techniques, this scheme does neither require precise intensity measurements nor information on the relation between intensity and shift. The residual relative variation of the stabilized laser frequency is only a third-order function of $|\Delta_L - \Delta_S|$ and can be kept small over a range of compensation mismatch. In order to fulfill this condition, a stabilization using the HRS scheme can be combined with a second servo system where Rabi spectroscopy with the same probe light intensity is used and the frequency offset from the HRS stabilization is determined. This offset can be used as an estimate of Δ_L in the HRS excitation so that the combined action of the two servo systems minimizes $|\Delta_L - \Delta_S|$ and ensures that slow variations of Δ_L will not degrade the light shift suppression. The experiments demonstrated a suppression of the light shift by four orders of magnitude and the predicted immunity against its fluctuations [17]. Under our present experimental conditions, we estimate the remaining uncertainty from the light shift as 1.1×10^{-18} [18].

The remaining largest shift of the transition frequency is caused by the Stark shift induced by the thermal blackbody radiation (BBR) emitted by the ion's environment. It is a significant advantage of trapped ion optical clocks that they generally have smaller BBR shifts than those based on neutral atoms [5]. This is primarily due to the blue shifting of the atomic resonances as a result of the tighter binding of the remaining electrons after ionization, thereby increasing the detuning between the thermal radiation and the atomic transitions. A complication in the determination of the BBR shift for trapped ions is that the strong radio frequency electric fields used to confine the ions will heat the dielectrics that are used as insulators in the trap structure and also leads to joule heating in the conductors. Heat removal from the trap is determined by conduction and radiation. It is therefore not appropriate to use the vacuum chamber temperature to estimate the BBR shift, as the trap structure subtends a large fraction of the solid angle visible to the ion and may be at a different temperature than the chamber. A combination of finite element modeling with measurements made with an infrared camera and temperature sensors at critical test points of a copy of the operational trap of the Yb^+ frequency standard at PTB has allowed us to determine the effective temperature rise of the radiation seen by the ion as 2.1(1.1) K [19]. An optimized trap design with an exclusive use of low-loss dielectrics and improved thermal contact from the trap electrodes to the vacuum feedthrough will allow one to reduce the temperature rise and uncertainty further [20].

Correction of the BBR shift requires knowledge of the temperature and of the differential polarizability between the ground and excited state, together with a small dynamic correction that must be applied due to the overlap of the thermal radiation spectrum with the atomic resonances [5]. For the Yb$^+$ E3 transition the problem is facilitated by the fact that all electric dipole transitions that are relevant for the differential Stark shift are at wavelengths smaller than 380 nm and have thus little overlap with the spectrum of thermal radiation at room temperature. A precise measurement of the differential polarizability can therefore be performed by measuring the light shift induced by near-infrared laser radiation. Lasers at 0.85 μm, 1 μm, 1.3 μm and 1.5 μm have been used and light shift profiles measured over the intensity distribution of the focused laser beams [21]. Combining the results of the thermal analysis of the trap [19] and the polarizability measurements, the BBR shift for the Yb$^+$ E3 transition is -45 mHz and the contribution to the relative uncertainty of the frequency standard 1.8×10^{-18} [18].

In total, the evaluation of the Yb$^+$ E3 optical clock now results in a systematic uncertainty of 3×10^{-18} for the realization of the unperturbed transition frequency, with the largest contribution now caused by the relativistic Doppler shift from residual motion of the ion [18]. Ensuing steps of further improvement are therefore foreseeable with a trap with lower motional heating rates, ground state cooling of the ion and with improved techniques for the cancellation of micromotion.

4. Remote frequency comparisons of optical clocks

There is much activity in the field of highly-stable optical frequency transfer using an optical carrier frequency in the range of 1.5 μm wavelength and transmission via single-mode optical fibers for telecommunication. It has been shown that with the high available bandwidth this method supports transfer of the full accuracy and stability of an optical clock over distances reaching 1000 km, like between Braunschweig, Munich and Paris [22, 23]. This technique is likely to find application for connections between metrology institutes, especially in densely populated regions of Europe and Asia with abundant broadband fiber infrastructure. For comparisons of optical clocks over intercontinental distances and on a global scale, the established methods of time and frequency transfer via satellites will be available, albeit at the restricted bandwidth of microwave links.

Two proof-of-principle measurement campaigns for frequency comparisons of optical clocks via satellites have recently been performed by consortia of national metrology institutes in Europe. For three weeks in June 2015, five optical clocks at NPL in the UK, LNE-SYRTE in France and PTB in Germany have been compared using two-way time and frequency transfer via a geostationary telecommunication satellite at a higher bandwidth, i.e. higher chiprate in the transmitted pseudo-random code of 20 Mchip/s instead of 1 Mchip/s used in routine operation for the comparison of Cs clocks. A link instability of 2.6×10^{-16} has been reached in 1 day of averaging time. At the time of writing, the data analysis for the actual optical frequency comparisons between the different institutes is still ongoing.

In the second campaign, GPS signals and geodetic post processing was used to perform a direct remote comparison of two optical frequency standards based on single laser-cooled ^{171}Yb$^+$ ions operated at NPL and PTB [24]. Precise Point Positioning (PPP) is a well-established GPS carrier-phase frequency transfer method [25, 26]. At both institutes an active hydrogen maser served as a flywheel oscillator that was connected to a GPS receiver as an external frequency reference and compared simultaneously to a realization of the unperturbed frequency of the E2 transition in ^{171}Yb$^+$ via an optical femtosecond frequency comb. For a total measurement time of 67 hours, a fractional frequency difference $y(\text{PTB}) - y(\text{NPL})$ of $-1.3(1.2) \times 10^{-15}$ was found, limited by the statistical uncertainty of the link and by the requirement to extrapolate over interruptions in the optical clock signals [24]. The result is consistent with an agreement of both optical clocks and with recent absolute frequency measurements made against caesium fountain clocks at both institutes.

5. Tests of fundamental physics

The new level of performance of the optical clocks will open opportunities for improved searches for violations of Einstein's equivalence principle that appear in models of grand unification and quantum gravity. The variety of different reference transitions offers sensitivities to a wider range of supposed novel interactions. The two reference transitions (E2 and E3) in Yb$^+$, for example, have largely different relativistic contributions to the excited state energies. A supposed relative change in the value of the fine structure constant α would therefore appear with an amplification factor of about 7 as a relative change in the frequency ratio of both transitions [27]. Using the record of precise frequency measurements of the Yb$^+$ E2 and E3 transitions against primary caesium clocks at PTB over several years, together with data from other precision frequency comparisons, we have recently constrained temporal variations of α and the proton-to-electron mass ratio $\mu = m_p/m_e$ more strictly: a least-squares analysis of all available experimental data yields $(1/\alpha)(d\alpha/dt) = -0.20(20) \times 10^{-16}/\text{yr}$ and $(1/\mu)(d\mu/dt) = -0.5(1.6) \times 10^{-16}/\text{yr}$ [28, 29]. When ratios of optical transition frequencies with an uncertainty of 10^{-18} become available over a period of several years, the sensitivity of this search will increase by about two orders of magnitude.

Acknowledgements

I would like to thank my colleagues at PTB who have contributed to the work that is summarized here: Chr. Tamm, B. Lipphardt, N. Huntemann, M. Okhapkin, Chr. Sanner, S. Weyers, V. Gerginov, M. Kazda, U. Sterr, C. Lisdat, F. Riehle, J. Leute, F. Riedel, E. Benkler, and our cooperation partners A. V. Taichenachev, V. I. Yudin, M. Doležal, P. Balling, M. Merimaa, and P. Gill. This work was partly funded by Deutsche Forschungsgemeinschaft within QUEST and by the EMRP project SIB04 Ion Clock. The EMRP is jointly funded by the EMRP participating countries within EURAMET and the European Union.

References

[1] H. Dehmelt, Coherent spectroscopy on single atomic system at rest in free space, *J. Phys. (Paris)* **42**, C8-299 (1981).

[2] C. W. Chou, D. B. Hume, J. C. J. Koelemeij, D. J. Wineland, and T. Rosenband, Frequency comparison of two high-accuracy Al$^+$ optical clocks, *Phys. Rev. Lett.* **104**, 070802 (2010).

[3] H. Katori, Spectroscopy of strontium atoms in the Lamb-Dicke confinement, in *Proc. of the 6th Symposium on Frequency Standards and Metrology* (World Scientific, Singapore, 2002).

[4] N. Poli, C. W. Oates, P. Gill and G. M. Tino, Optical atomic clocks, *Riv. Nuovo Cim.* **36**, 555 (2013).

[5] A. D. Ludlow, M. M. Boyd, Jun Ye, E. Peik, and P. O. Schmidt, Optical atomic clocks, *Rev. Mod. Phys.* **87**, 637 (2015).

[6] K. Numata, A. Kemery, and J. Camp, Thermal-noise limit in the frequency stabilization of lasers with rigid cavities, *Phys. Rev. Lett.* **93**, 250602 (2004).

[7] T. Kessler, C. Hagemann, C. Grebing, T. Legero, U. Sterr, F. Riehle, M. J. Martin, L. Chen, and J. Ye, A sub-40 mHz linewidth laser based on a silicon single-crystal optical cavity, *Nat. Photonics* **6**, 687 (2012).

[8] S. Häfner, S. Falke, Chr. Grebing, S. Vogt, Th. Legero, M. Merimaa, Chr. Lisdat, and U. Sterr, 8×10^{17} fractional laser frequency instability with a long room-temperature cavity, *Opt. Lett.* **40**, 2112 (2015).

[9] G. D. Cole, Wei Zhang, M. J. Martin, Jun Ye, and M. Aspelmeyer, Tenfold reduction of Brownian noise in optical interferometry, *Nat. Photonics* **7**, 644 (2013).

[10] Chr. Tamm, S. Weyers, B. Lipphardt, and E. Peik, Stray-field-induced quadrupole shift and absolute frequency of the 688-THz ^{171}Yb$^+$ single-ion optical frequency standard, *Phys. Rev. A* **80**, 043403 (2009).

[11] Chr. Tamm, D. Engelke, and V. Bühner, Spectroscopy of the electric-quadrupole transition $^2S_{1/2} \to {}^2D_{3/2}$ in trapped ^{171}Yb$^+$, *Phys. Rev. A* **61**, 053405 (2000).

[12] M. Roberts, P. Taylor, G. P. Barwood, P. Gill, H. A. Klein, and W. R. C. Rowley, Observation of an electric octupole transition in a single ion, *Phys. Rev. Lett.* **78**, 1876 (1997).

[13] S. A. Webster, P. Taylor, M. Roberts, G. P. Barwood, and P. Gill, Kilohertz-resolution spectroscopy of the $^2S_{1/2} - {}^2 F_{7/2}$ electric octupole transition in a single ^{171}Yb$^+$ ion, *Phys. Rev. A* **65**, 052501 (2002).

[14] N. Huntemann, M. V. Okhapkin, B. Lipphardt, Chr. Tamm, and E. Peik, High-accuracy optical clock based on the octupole transition in ^{171}Yb$^+$, *Phys. Rev. Lett.* **108**, 090801 (2012).

[15] S. N. Lea, Limits to time variation of fundamental constants from comparisons of atomic frequency standards, *Rep. Prog. Phys* **70**, 1473 (2007).

[16] V. I. Yudin, A. V. Taichenachev, C. W. Oates, Z. W. Barber, N. D. Lemke, A. D. Ludlow, U. Sterr, C. Lisdat, and F. Riehle, Hyper-Ramsey spectroscopy of optical clock transitions, *Phys. Rev. A* **82**, 011804 (2010).

[17] N. Huntemann, B. Lipphardt, M. V. Okhapkin, Chr. Tamm, E. Peik, A. V. Taichenachev, V. I. Yudin, Generalized Ramsey excitation scheme with suppressed light shift, *Phys. Rev. Lett.* **109**, 213002 (2012).

[18] N. Huntemann et al., *to be published*.

[19] M. Doležal, P. Balling, P. B. R. Nisbet-Jones, S. A. King, J. M. Jones, H. A. Klein, P. Gill, T. Lindvall, A. E. Wallin, M. Merimaa, C. Tamm, C. Sanner, N. Huntemann, N. Scharnhorst, I. D. Leroux, P. O. Schmidt, T. Burgermeister, T. E. Mehlstäubler and E. Peik, Analysis of thermal radiation in ion traps for optical frequency standards, *Metrologia, in print*, arXiv:1510.05556 (2015).

[20] P. B. R. Nisbet-Jones, S. A. King, J. M. Jones, R. M. Godun, C. F. A. Baynham, K. Bongs, M. Doležal, P. Balling, and P. Gill, A single-ion trap with minimized ion-environment interactions, arXiv:1510.06341 (2015).

[21] N. Huntemann, High-accuracy optical clock based on the octupole transition in ^{171}Yb$^+$, *PhD Thesis* (Leibniz-Universität Hannover, 2014).

[22] K. Predehl, G. Grosche, S. M. F. Raupach, S. Droste, O. Terra, J. Alnis, Th. Legero, T. W. Hänsch, Th. Udem, R. Holzwarth, and H. Schnatz, A 920-kilometer optical fiber link for frequency metrology at the 19th decimal place, *Science* **336**, 441 (2012).

[23] S. M. F. Raupach, A. Koczwara, and G. Grosche, Brillouin amplification supports 1×10^{-20} uncertainty in optical frequency transfer over 1400 km of underground fiber, *Phys. Rev. A* **92**, 021801(R) (2015).

[24] J. Leute, N. Huntemann, B. Lipphardt, Chr. Tamm, P. B. R. Nisbet-Jones, S. A. King, R. M. Godun, J. M. Jones, H. S. Margolis, P. B. Whibberley, A. Wallin, M. Merimaa, P. Gill, and E. Peik, Frequency comparison of ^{171}Yb$^+$ ion optical clocks at PTB and NPL via GPS PPP, arXiv:1507.04754 (2015).

[25] J. Kouba and P. Héroux, Precise point positioning using IGS orbit and clock products, *GPS Solutions* **5**, 12 (2001).

[26] J. M. Dow, R. E. Neilan, and C. Rizos, The international GNSS service in a changing landscape of Global Navigation Satellite Systems, *Journal of Geodesy*, **83**, 191 (2009).

[27] V. V. Flambaum and V. A. Dzuba, Search for variation of the fundamental constants in atomic, molecular, and nuclear spectra, *Can. J. Phys.* **87**, 25 (2009).

[28] N. Huntemann, B. Lipphardt, Chr. Tamm, V. Gerginov, S. Weyers, and E. Peik, Improved limit on a temporal variation of m_p/m_e from comparisons of Yb$^+$ and Cs atomic clocks, *Phys. Rev. Lett.* **113**, 210802 (2014).

[29] R. Godun, P. Nisbet-Jones, J. Jones, S. King, L. Johnson, H. Margolis, K. Szymaniec, S. Lea, K. Bongs, and P. Gill, Frequency ratio of two optical clock transitions in ^{171}Yb$^+$ and constraints on the time variation of fundamental constants. *Phys. Rev. Lett.* **113**, 210801 (2014).

Precision Measurement of the Newtonian Gravitational Constant by Atom Interferometry

G. Rosi*, G. D'Amico and G. M. Tino

Dipartimento di Fisica e Astronomia & LENS, Università di Firenze, INFN Sezione di Firenze, via Sansone 1, I-50019 Sesto Fiorentino (FI), Italy

L. Cacciapuoti

European Space Agency, Keplerlaan 1, 2201 AZ Noordwijk, The Netherlands

M. Prevedelli

Dipartimento di Fisica dell'Università di Bologna, Via Irnerio 46, I-40126, Bologna, Italy

F. Sorrentino

INFN Sezione di Genova, Via Dodecaneso 33, I-16146 Genova, Italy
E-mail: nguros@tin.it

We report on the latest determination of the Newtonian gravitational constant G using our atom interferometry gravity gradiometer. After a short introduction on the G measurement issue we will provide a description of the experimental method employed, followed by a discussion of the experimental results in terms of sensitivity and systematic effects. Finally, prospects for future cold atom-based experiments devoted to the measurement of this fundamental constant are reported.

Keywords: Atom Interferometry, Metrology, Gravity, Newtonian Constant.

1. Introduction

Nowadays most of the physical constants are known within a few parts per billion, in the worst cases some parts per million. One of the few exceptions is the gravitational constant G, introduced for the first time by Newton in 1665 to describe the attractive force between all bodies with mass. Despite

$$\mathbf{F} = -G\frac{m_1 m_2}{r^3}\mathbf{r} \qquad (1)$$

being one of the best known among all physical laws, the CODATA-recommended values of G in the last decade have a relative uncertainty in the range of 100 ppm, recently reduced to 47 ppm [1]. Besides the purely metrological interest, there are several reasons why a more precise determination of G is important: in astronomy, the factor GM of astronomical objects can be determined extremely well and thus a better knowledge of G leads to a better knowledge of M, which in turn leads to a better physical understanding of celestial bodies; in geophysics, uncertainties of density and elastic parameters of the Earth are directly related to the uncertainties on G [2]; in theoretical physics, spatial and temporal variations of G are predicted by some theories (see [3] for a review on this topic). The reasons for the difficulties in the determination of G can be found in the peculiar nature of the gravitational force: gravity cannot be shielded and its weakness allows other forces to contribute with big systematic effects in laboratory experiments [4]. Secondly, the majority of the experiments performed so far are based on macroscopic suspended masses: parasitic couplings in suspending fibers are not well understood and could be responsible for the observed discrepancies. Instead, using microscopic masses such as neutral atoms to probe the gravitational field generated by a well characterized source mass can solve this kind of problems. For all these reasons, atom interferometry represents an alternative and powerful method.

Matter-wave interferometers are newcomers for experimental gravitation but already they have been successfully employed to measure gravitational acceleration [6, 7] and gravity gradients [8, 11], local gravity curvature [12], as gyroscopes based on the Sagnac effect [13], for testing the $1/r^2$ law [9], and in applications in geophysics [10]. Proof-of-principle experiments to measure G using atom interferometry [14] have been reported at various levels of accuracy: 10000 ppm in Florence [15], 5000 ppm in Stanford [16] and again 2000 ppm in Florence [17]. Finally, in 2014, we reported an uncertainty of 150ppm [18] which is for the first time comparable with that of the CODATA value at that time [19].

Here we are going to revise this last G determination in all its most relevant aspects. The manuscript is organized as follow: in section 2 we illustrate the principle of measurement and the experimental procedure; in section 3 we present the ultimate sensitivity of our instrument and the experimental data set used to determine the phase shift induced by our source masses; in section 4 we describe the numerical method adopted to estimate the G value from raw data, showing also the final error budget with a brief discussion of the major sources of systematic errors; finally, in section 5, prospects for future atom interferometry-based G determinations are presented.

2. Apparatus and experimental protocol

As already mentioned, one of the main issue in the determination of G depends on the presence of unidentified systematic errors, due to the intrinsic weakness of gravity with respect to other forces. In order to effectively isolate the gravitational signal, our experiment is designed with a double-differential configuration: the atomic sensor is a double interferometer in a gravity gradiometer configuration, to subtract common-mode spurious signals, and to produce the gravitational field we used two sets of well-characterized tungsten masses that were placed in two different positions to modulate the relevant gravitational signal. The basic ingredient of this instrument lies in the realization of the so-called Mach-Zehnder interferometer [5]. Such kind of scheme is achieved using a $\pi/2 - \pi - \pi/2$ sequence of three Raman pulses separated by time intervals T in counter-propagating configuration

[6] that couples the two hyperfine ground states of an alkali atom (in our case ^{87}Rb) and trasfers at the same time a momentum $\hbar k_{\text{eff}}$, where k_{eff} is the sum of the Raman lasers wave vectors. In this picture the $\pi/2$ and π pulses realize respectively the beam-splitters and mirrors of the interferometer. At the interferometer output, the probability of detecting the atoms in the initial internal state $|a\rangle$ is given by $P_a = (1 + \cos(\phi))/2$, where ϕ represents the phase difference accumulated by the wave packets along the two interferometer arms. This phase shift can be directly connected to the gravity acceleration experienced by the atomic wave-packet. The gravity gradiometer consists of two absolute accelerometers operated in differential mode. Two spatially separated atomic clouds in free fall along the same vertical axis are simultaneously interrogated by the same Raman beams to provide a measurement of the differential acceleration induced by gravity on the two samples. A Lissajous figure results from the composition of the P_a trace of the upper accelerometer versus the lower one [24]. The differential phase shift $\Phi = \phi_u - \phi_d$, which is proportional to the gravity gradient, is obtained from the eccentricity and the rotation angle of the ellipse best fitting the experimental data. The scheme of our interferometer is shown in Figure 1 for two different configurations of the source masses, together with a picture of the apparatus. Two cold ^{87}Rb atomic samples (T $\sim 3~\mu$K) are launched in the interferometry region separated ~ 30 cm by juggling the atoms loaded in a 3D-MOT [20]. Before realizing the interferometer, a series of velocity selection and blow-away pulses are employed in order to prepare the atoms into the magnetically insensitive $m_F = 0$ sub-level of the $5^2S_{1/2}$ hyperfine ground state and in a well-defined velocity class. The interferometer sequence takes place around the center of the vertical tube shown in Fig 1. In this region, surrounded by two μ-metal shields (76 dB attenuation factor of the magnetic field in the axial direction), a uniform magnetic field of 25 μT along the vertical direction defines the quantization axis. The field gradient along this axis is lower than 5 μT/m. After the Raman interferometry sequence, the normalized population of the ground state is measured in a chamber placed just above the MOT by selectively exciting the atoms on the $F = 1, 2$ hyperfine levels and detecting the resulting fluorescence. In order to modulate the actual value of the gravity gradient or, more properly, the gravity acceleration difference felt by the two atomic samples, source masses are vertically displaced in two different configurations, in order to reduce it (configuration C_2) or enhance it (configuration C_1). Knowing all the geometric parameters of the atomic sample and source masses distribution, the value of G can be determined, being directly proportional to the differential angle

$$\Delta\Phi = \Phi_{C_1} - \Phi_{C_2}. \tag{2}$$

This value can indeed be numerically evaluated [21], leaving in this way G as the unique free parameter. The source masses [22] are composed of 24 tungsten alloy (INERMET IT180) cylinders, for a total mass of about 516 kg. They are positioned on two titanium platforms and distributed in hexagonal symmetry around the vertical axis of the tube. Shape and position of each cylinder must be determined with

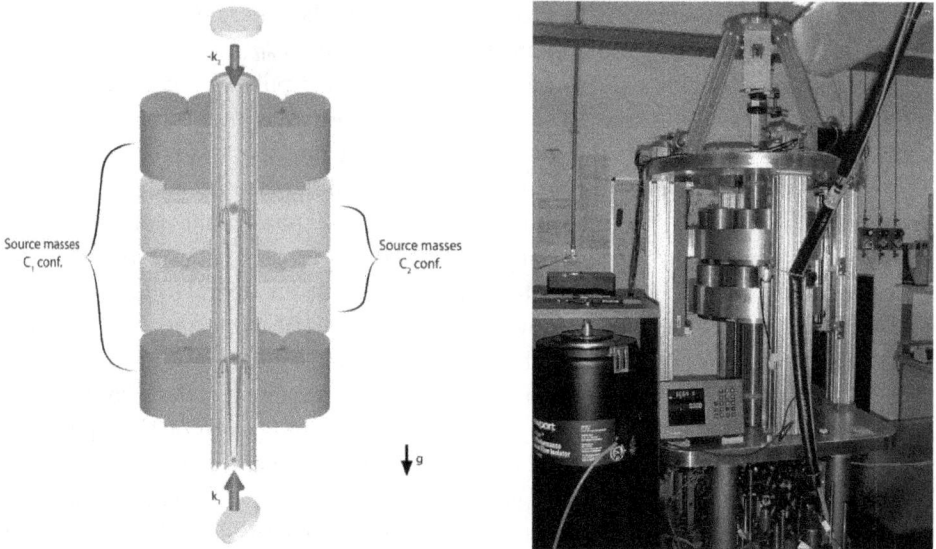

Fig. 1. Left: scheme of the gravity gradiometer. ^{87}Rb atoms, trapped and cooled in a magneto-optical trap (MOT), are launched upwards in a vertical vacuum tube with a moving optical molasses scheme, producing an atomic fountain. Near the apogees of the atomic trajectories, a measurement of their vertical acceleration is performed by a Raman interferometry scheme. External source masses are positioned in two different configurations (C_1 and C_2) and the induced phase shift is measured as a function of masses positions. Right: a picture of the apparatus for the atomic fountain, including the source masses for G measurement.

micrometric accuracy, in order to reduce the systematic uncertainty. Consequently all the cylinders have been machined and polished in order to regularize the shape. Additionally, a conical hole was placed in the center of each cylinder base to easily perform position evaluation using small steel spheres. We investigated the shape of each cylinder with a contact 3D scanner (Brown & Sharpe DEA Scirocco) able to perform position measurements with 1 μm accuracy. To perform accurate mass measurements we decided to compare each tungsten cylinder with a sample mass (21500 g) characterized at the milligram level. We used a precision balance with a 1 mg resolution (courtesy of INRIM). Then, the cylinders have been placed on the two holding platforms, trying to get the most regular arrangement with the help of a caliper. Afterward, taking conical holes as measuring points, the 3D position of each cylinder was evaluated using a laser tracker with an effective resolution of 10 μm.

3. Gradiometer sensitivity and G measurement

One of the key points in doing precision measurements is to have an instrument with the highest level of sensitivity, in order to efficiently detect systematic effects. In this spirit several efforts were made in order to analyze the influence of the most relevant experimental parameters on the stability and accuracy of our gravity gradiometer. We actively stabilize most of them, i.e. the optical intensity of cooling, Raman and probe laser beams, acting on the RF power driving acousto-optical modulators, and the Raman mirror tilt, acting on a piezo tip/tilt system [23]. The servo on cooling and Raman lasers intensity is implemented by means of a slow digital loop, sampling the four powers (up and down cooling beams, master and slave Raman beams) every 72 experimental cycles ($\sim$30 minutes) and driving the RF power corresponding AOMs through a numerical loop filter. Residual fluctuations are below 0.3 %. The active control of cooling, Raman and probe laser intensities, together with Coriolis compensation (see next section for details), improves the long term stability of differential gravity measurements considerably. Several other improvements of the apparatus have allowed us to increase the number of atoms and the repetition rate of the experiment, and also to reduce the technical noise at detection and increase the ellipse contrast. Figure 2 shows the Allan deviation of the experimental data up

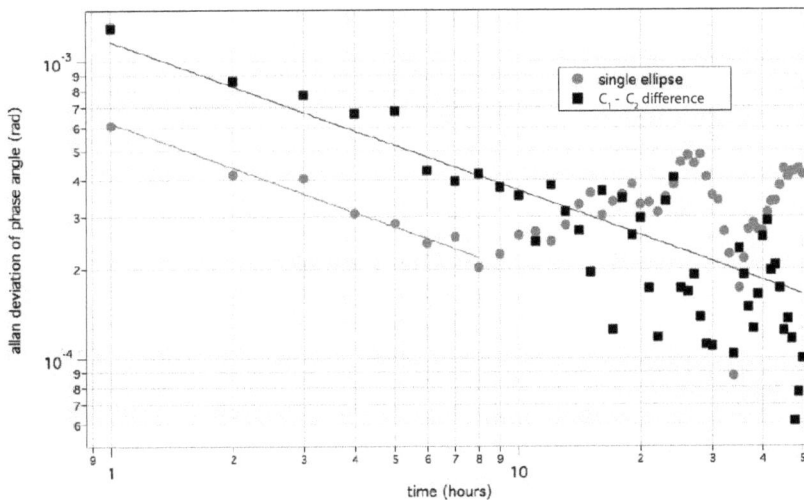

Fig. 2. Allan deviation of the atom interferometry phase; circles represent the measurement of gravity gradient for a given position of the source masses; squares represent the diffential phase measurement for two configurations of source masses [25].

to an integration time of 50 hours, for a fixed configuration of source masses and for the difference of measurements in the two configurations. We currently achieve a sensitivity of 13 mrad at 1 s, in agreement with the calculated quantum projection noise (QPN) limit for 2×10^5 atoms, corresponding to a sensitivity to differential accelerations of 3×10^{-9} g at 1 s.

The value of the Newtonian gravitational constant was obtained from a series of gravity gradient measurements performed by periodically changing the vertical

66

position of the source masses between the two configurations every 30 minutes and acquiring 720 data points. Figure 3 shows the data used for the determination of G (right part), together with tipical Lissajous plots (left part). Data were collected in 100 hours during one week in July 2013. The resulting value of the differential phase shift $\Delta\Phi$ is 0.547870(63) rad.

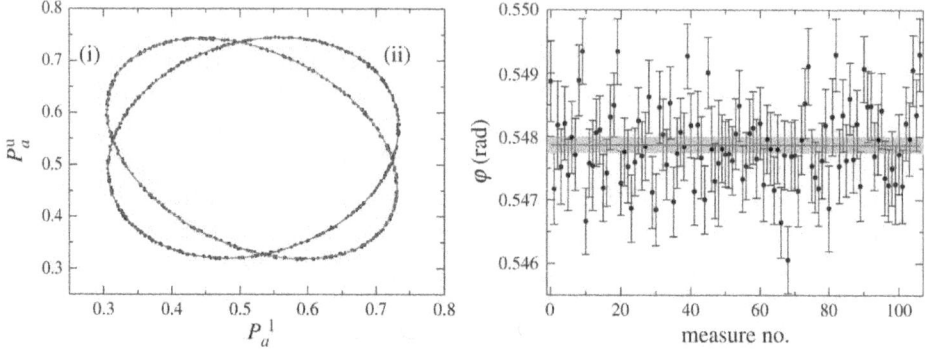

Fig. 3. Left: Two ellipses corresponding to the C_2 (i) and C_1 (ii) source masses configuration. The continuous lines are the best fits to the experimental points. Right: Results of the measurements to determine G. Data acquisition for each point took about one hour [12].

4. Numerical simulation and systematic effects

The numerical simulation of the experiment plays a central role in evaluating G from raw data. Basically it provides a simulated phase Φ_s that can be written as $\Phi_s = \Xi G$, where the proportionality factor Ξ depends essentially on the geometrical features of the experiment. Therefore, in order to obtain G, it will be sufficient to match Φ_s with the measured Φ angle. The numerical simulation evaluates the gravitational potential produced by a given configuration of the source masses, implements the calculation of the phase shift experienced by single atoms at the two interferometers, and finally runs a Monte Carlo simulation on the atomic trajectories by varying the initial position and velocity out of the density and velocity distribution of the atomic cloud that best fits the experimental density profiles. The gravitational potential generated by the source masses is computed analytically using a multipole expansion. The phase shift is calculated using a perturbative method [21]: the Lagrangian L of the system can be separated into an unperturbed part L_0, which contains the kinetic energy and the linear part of earth's gravity potential, and a perturbative term L_1 which accounts for the contributions of both the Earths gravity gradient and the source masses. The phase shift produced by L_1 is equal to

$$\phi_{pert} = \frac{1}{\hbar} \int_{\Gamma_0} L_1 dt, \tag{3}$$

where Γ_0 is the unperturbed classical particle path. In order to evaluate systematic uncertainties and shifts, derivatives are computed with respect to all the input parameters, including the positions of the source masses. In Table 1 the resulting error budget is reported.

Table 1.   Corrections and uncertainties on Φ and propagated effect on G [18].

Effect	Uncertaintly	Correction to G (ppm)	Relative uncertainty $\Delta G/G$ (ppm)
Air density	10%	60	6
Apogee time	30 μs		6
Clouds horizontal size	0.5 mm		24
Clouds vertical size	0.1 mm		56
Clouds horizontal position	1 mm		37
Clouds vertical position	0.1 mm		5
Launch direction change	8 μrad		36
Cylinders density inhomogenity	10^{-4}	91	18
Cylinders radial position	10 μm		38
Ellipse fitting		-13	4
Size of detection region	1 mm		13
Support platforms mass	10 g		5
Translation stages position	0.5 mm		6
Other effects		< 2	1

It is worth to mention that not all the entries were evaluated through the simulation. In particular the effect on Φ of the launch direction change via Coriolis effect was evaluated by a direct phase shift measurement. In the laboratory reference frame the Coriolis acceleration adds to the single gradiometer a phase shift term

$$\Phi = 2\mathbf{k}_{\text{eff}} \cdot (\mathbf{\Omega} \times \mathbf{\Delta v})T^2, \tag{4}$$

where $\mathbf{\Omega}$ is its angular velocity due to the Earth rotation and $\mathbf{\Delta v}$ is the clouds velocity difference. $\mathbf{\Omega}$ can be varied acting on the retroreflecting Raman mirror with a tip-tilt piezo system and thus possible variations of $\mathbf{\Delta v}$ when moving the masses between the Close and Far configurations can be characterized and compensated [26].

A more complete description of each systematic shift entry presented in Table 1 can be found in [27].

As a final result we obtain the value $G = 6.67191(77)(65) \times 10^{-11}$ m^3 kg^{-1} s^{-2}. The statistical and systematic errors, reported in parenthesis as one standard deviation, lead to a combined relative uncertainty of 150 ppm.

5. Outlook and conclusions

Pushing the accuracy on G towards the 10^{-5} level will require a tenfold improvement of both statistical and systematic uncertainties. Regarding the first issue, a standard but quite tedious method consists in increasing the atomic flux and/or decreasing the cycle time of the experiment. All these improvements will reduce the relative uncertainty on Φ as $1/\sqrt{N}$ where N is the number of detected atoms. Trying to improve towards the $1/N$ Heisenberg limit is an active field of research and several techniques have been developed [28, 29] but up to now a strategy to implement them in an atom interferometer has not been identified.

Reducing the systematics will obviously require a tighter control on some experimental parameters, in particular: cloud sizes and trajectories, compensation of the Earth rotation and density inhomogeneity for the source masses. Regarding the first task, using colder atoms and smaller clouds as implemented in [30] appears a promising approach, while the accuracy of the tip tilt system adopted to compensate the Coriolis effect can be easily improved by an order of magnitude. The problem of density inhomogeneity can be fixed adopting different materials, for example silicon. In this case, however, a decrement in the source masses density value and thus in the differential phase shift is expected. To overcome this limitation large momentum trasfer interferometers [31] will allow to obtain the same signal and signal-to-noise ratio while reducing the source masses.

In conclusion a precision measurement of G with a gradiometer based on cold atoms was performed, obtaining an uncertainty in the 100 ppm range. We are also confident that, in the near future, the accuracy of such method will be brought in the 10 ppm range. This can help to solve the G puzzle.

References

[1] P. J. Mohr, D. B. Newell and B. N. Taylor, "CODATA Recommended values of the fundamental physical constants: 2014", *arXiv:1507.07956 [physics.atom-ph]* (2015)

[2] W. Torge, "Gravimetry", De Gruyter, New York, (1989)

[3] G. T. Gillies, "The Newtonian gravitational constant: recent measurements and related studies", *Rep. Prog. Phys.* **60** 151225 (1997)

[4] T. Quinn, "The Newtonian constant of gravitation, a constant too difficult to measure?" *Phil. Trans. R. Soc. A* **372** 20140286 (2014)

[5] G. M. Tino and M. Kasevich, *Proc. Int. School Phys. "Enrico Fermi"*, Course CLXXXVIII, Varenna 2013 (Società Italiana di Fisica and IOS Press), (2014)

[6] M. Kasevich and S. Chu, "Atomic interferometry using stimulated Raman transitions", *Phys. Rev. Lett.* **67** 181-184 (1991)

[7] A. Peters, K. Y. Chung and S. Chu, "Measurement of gravitational acceleration by dropping atoms", *Nature* **400** 849-852 (1999)

[8] J. M. McGuirk, G. T. Foster, J. B. Fixler, M. J. Snadden and M. A. Kasevich, "Sensitive absolute-gravity gradiometry using atom interferometry", *Phys. Rev. A* **65** 033608 (2002)

[9] F. Sorrentino, A. Alberti, G. Ferrari, V. V. Ivanov, N. Poli, M. Schioppo and G. M. Tino, "Quantum sensor for atom-surface interactions below 10 μm", *Phys. Rev. A* **79** 013409 (2009)

[10] M. de Angelis, A. Bertoldi, L. Cacciapuoti, A. Giorgini, G. Lamporesi, M. Prevedelli, G. Saccorotti, F. Sorrentino and G. M. Tino, "Precision gravimetry with atomic sensors", *Meas. Sci. Technol.* **20** 022001 (2009)

[11] F. Sorrentino, Y.-H. Lien, G. Rosi, L. Cacciapuoti, M. Prevedelli and G. M. Tino, "Sensitive gravity-gradiometry with atom interferometry: Progress towards an improved determination of the gravitational constant", *New J. Phys.* **12** 095009 (2010)

[12] G. Rosi, L. Cacciapuoti, F. Sorrentino, M. Menchetti, M. Prevedelli and G.M. Tino, "Measurement of the gravity-field curvature by atom interferometry", *Phys. Rev. Lett.* **114** 013001 (2015)

[13] A. Gauguet, B. Canuel, T. Lvque, W. Chaibi and A. Landragin, "Characterization and limits of a cold-atom Sagnac interferometer", *Phys. Rev. A* **80** 063604 (2009)

[14] M. Fattori, G. Lamporesi, T. Petekski, J. Stuhler and G.M. Tino, "Towards an atom interferometric determination of the newtonian gravitational constant", *Phis. Lett. A* **318** 184 (2003)

[15] A. Bertoldi, G. Lamporesi, L. Cacciapuoti, M. de Angelis, M. Fattori, T. Petelski, A. Peters, M. Prevedelli, J. Stuhler and G. M. Tino, "Atom interferometry gravity-gradiometer for the determination of the Newtonian gravitational constant *G*", *Eur. Phys. J. D* **88** 271279 (2006)

[16] J. B. Fixler, G. T. Foster, J. M. McGuirk and Kasevich, "Atom interferometer measurement of the newtonian constant of gravity", *Science* **315** 74-77 (2007)

[17] G. Lamporesi, A. Bertoldi, L. Cacciapuoti, M. Prevedelli and G. M. Tino, "Determination of the Newtonian Gravitational Constant Using Atom Interferometry", *Phys. Rev. Lett.* **100** 050801 (2008)

[18] G. Rosi, F. Sorrentino, L. Cacciapuoti, M. Prevedelli and G. M. Tino, "Precision measurement of the Newtonian gravitational constant using cold atoms", *Nature* **510** 518-521, (2014)

[19] P. J. Mohr, Barry N. Taylor and David B. Newell, "CODATA recommended values of the fundamental physical constants:2010" *Rev. Mod. Phys* **84** 1527, (2012)

[20] R. Legere and K. Gibble, "Quantum scattering in a juggling atomic fountain", *Phys. Rev. Lett.* **81** 5780 (1998)

[21] P. Storey and C. Cohen-Tannoudji, "The Feynman path integral approach to atomic interferometry. A tutorial", *J. Phys. II France* **4** 1999-2027 (1994)

[22] G. Lamporesi, A. Bertoldi, A. Cecchetti, B. Dulach, M. Fattori, A. Malengo, S. Pettorruso, M. Prevedelli and G. M. Tino, "Source mass and positioning system for an accurate measurement of *G*", *Rev. Sci. Instrum.* **78** 075109 (2007)

[23] J. M. Hogan, D. M. S. Johnson and M. A. Kasevich, "Proceedings of the International School of Physics Enrico Fermi Course CLXVIII on Atom Optics and Space Physics", *IOS Press, Oxford* (2007)

[24] G. T. Foster, J. B. Fixler, J. M. McGuirk and M. Kasevich, "Method of phase extraction between coupled atom interferometers using ellipse-specific fitting", *Optics Letters* **27** 951-953 (2002)

[25] G.M. Tino, G. Rosi, F. Sorrentino, L. Cacciapuoti and M. Prevedelli, "High precision measurement of the gravitational constant with atom interferometry", *European Frequency and Time Forum & International Frequency Control Symposium (EFTF/IFC)* 593-598 (2013)

[26] G. Rosi, "Precision gravity measurements with atom interferometry", *Ph.D. thesis, Università di Pisa* (2012)

[27] M. Prevedelli, L. Cacciapuoti, G. Rosi, F. Sorrentino and G. M. Tino, "Measuring the Newtonian constant of gravitation G with an atomic interferometer", *Phil. Trans. R. Soc. A* **372** 20140030 (2014)

[28] C.F. Ockeloen, R. Schmied, M. F. Riedel and P. Treutlein, "Quantum metrology with a scanning probe atom interferometer", *Phys. Rev. Lett.* **111** 143001 (2013)

[29] H. Zhang, R. McConnell, S. Uk, Q. Lin, M. H. Schleier-Smith, I. D. Leroux and V. Vuletić, "Collective state measurement of mesoscopic ensembles with single-atom resolution", *Phys. Rev. Lett.* **109** 133603 (2012)

[30] S. M. Dickerson, J. M. Hogan, A. Sugarbaker, D. M. S. Johnson and M. A. Kasevich, "Multiaxis inertial sensing with long-time point source atom interferometry", *Phys. Rev. Lett.* **111** 083001 (2013)

[31] G. D. McDonald, C. C. N. Kuhn, S. Bennetts, J. E. Debs, K. S. Hardman, M. Johnsson, J. D. Close and N. P. Robins, "$80\hbar k$ momentum separation with Bloch oscillations in an optically guided atom interferometer", *Phys. Rev. A* **88** 053620 (2013)

Optical Sideband Cooling of Ions in a Penning Trap

R. C. Thompson, J. F. Goodwin, G. Stutter and D. M. Segal (deceased 23 September 2015)

Department of Physics, Imperial College London,
London SW7 2AZ, UK
* E-mail: r.thompson@imperial.ac.uk
www3.imperial.ac.uk/iontrap

Optical sideband cooling is a well-established technique for preparing trapped ions in the ground state of one or more of their motional degrees of freedom. Up to now, this technique has mainly been applied in various types of radiofrequency trap. We show in this paper that the technique is also very effective in a Penning trap, and demonstrate that our trap has an exceptionally low heating rate, due to the large size of the Penning trap electrodes and the absence of large-amplitude radiofrequency fields in the trap.

Keywords: Penning trap, laser cooling, ion Coulomb crystals.

1. Introduction

Optical sideband cooling of a single trapped ion in an RF trap was first demonstrated more than twenty years ago [1] and has since become well established as a standard technique in research using cold trapped ions. However, this technique has up to now not been demonstrated for an ion in a Penning trap. There are several reasons for this, the most important of which arises from the effect of the strong magnetic field on the energy levels of the trapped ion.

The Zeeman splitting arising from the magnetic field in a Penning trap is typically tens of GHz and this means that separate lasers are required to address transitions from each of the ground state Zeeman sublevels. In an ion such as Mg^+ or Be^+, where there is a closed $S_{1/2}$–$P_{3/2}$ laser cooling transition, it is possible to use optical pumping techniques to isolate a closed transition for Doppler laser cooling (see, for example, [2–5]). The Be^+ system can be used for quantum information studies, using the ground state hyperfine splitting as a qubit which can be addressed using microwaves or a Raman transition [6]. For work with an optical qubit, it is necessary to use an ion that has a metastable state accessible from the ground state of the ion. An example of such an ion is Ca^+. In this case it will always be possible for the ion to decay into a metastable electronic level from the excited P state, so now no simple closed laser cooling cycle is possible and multiple repump lasers are required (see Section 4). This greatly complicates the process of laser cooling.

In this article, we present a study of optical sideband spectroscopy and sideband cooling of an ion in a Penning trap for the first time. We demonstrate by a direct measurement that the Doppler limit is achieved for the axial motion, and we show that the equilibrium temperatures of the two radial motions are very different, with the magnetron motion cooled to well below the standard Doppler limit. We then show that optical sideband cooling can be applied to the axial motion in order to cool the ion to its ground state with high probability, and by measuring the heating rate of the axial motion we demonstrate that heating processes are much weaker in our trap than in radiofrequency traps. This is mainly due to the large size of the Penning trap electrodes.

This article is structured as follows. In Sections 2 and 3 we give a brief introduction to the physics of the Penning trap and to laser cooling in this system. Section 4 describes our experimental arrangement and Section 5 presents our work on spectroscopy and sideband cooling of the axial motion. In Section 6 we give a brief discussion of measurements of the radial motion. In Section 7 we show the results of experiments on ion Coulomb crystals in the Penning trap, and this is followed by some conclusions in Section 8.

2. The Penning trap

The Penning trap is an excellent device for confining atomic ions because the confinement is provided solely by static fields and the ions are well isolated from the environment. Axial confinement is provided by a positive potential on the endcaps, giving rise to axial oscillations at an angular frequency of $\omega_z = (4eV/mR^2)^{1/2}$ where e and m are the charge and mass of the ions, V is the applied voltage, and R is a constant related to the dimensions of the trap [7]. Radial confinement is achieved by the application of a static magnetic field along the axis of the trap, and gives rise to a motion which is a superposition of two radial oscillation frequencies. The first, ω_c', is called the modified cyclotron frequency and is equal to $\omega_c/2 + \omega_1$ where $\omega_c = eB/m$ is the pure cyclotron frequency corresponding to the magnetic field B, and ω_1 is related to the cyclotron and axial frequencies through

$$\omega_1^2 = \omega_c^2/4 - \omega_z^2/2. \tag{1}$$

The other radial frequency is called the magnetron frequency and is given by $\omega_m = \omega_c/2 - \omega_1$. The total energy of the magnetron mode is negative due to the fact that the radial potential energy is negative. This makes the magnetron motion unstable, however, the ions may nevertheless be confined for long periods under ultrahigh vacuum (UHV) conditions.

Penning traps have been used for a wide range of experiments in precision measurement, plasma physics, atomic physics, quantum optics and for the study of ion Coulomb crystals (see for example [6, 8–10]).

3. Laser cooling

Doppler laser cooling can be used in a Penning trap to reduce the temperature of a single trapped ion to the mK region. However, care has to be taken because the unstable nature of the magnetron motion means that it is heated rather than cooled if a red-detuned radial cooling beam is simply directed through the centre of the trap. This can be overcome by offsetting the laser beam from the centre of the trap so that there is a gradient of laser intensity at the position of the ion [11, 12], or by applying the axialisation technique [13]. Alternatively, a rotating wall potential can be applied to the trap when more than one ion is present [14, 15]. The general effect of these techniques is to supply angular momentum to the system, forcing the orbits of the ions to become smaller, thus reducing the kinetic energy of the system.

In our experiment we use conventional radial and axial laser cooling. However, as mentioned above, the radial laser beam has to be offset from the centre of the Penning trap in order to cool all degrees of freedom of the ions. For small numbers of ions, the offset is comparable to the width of the laser beam. We sometimes apply the 'axialisation technique', using a small radial quadrupole field which couples the two radial motions in order to improve the cooling of the magnetron motion. Using this technique also makes Doppler cooling of the ion more stable and reproducible.

The Doppler limit for calcium ions is approximately 0.5 mK. Although this limit applies to the axial motion, it is modified for the radial motions: the limit for the cyclotron motion is expected to be slightly higher, and that for the magnetron motion is expected to be much lower, in the region of tens of μK. This reflects the relatively low cooling rate for the magnetron motion [11].

In order to cool lower than the Doppler limit it is necessary to apply sideband cooling using a highly-stabilised laser to address a narrow optical transition. The ion motion induces sidebands which are resolved if the natural width of the transition and the laser linewidth are both small compared to the ion oscillation frequency. The laser is generally tuned to the first sideband on the lower frequency (red) side, so the vibrational quantum number is reduced by one each time an excitation takes place. Sideband cooling in the Penning trap works in exactly the same way as in radiofrequency traps, except that the relatively low oscillation frequencies in the Penning trap lead to high initial average quantum numbers, which implies that cooling will take longer. After sideband cooling the ion is expected to be in the ground state most of the time, limited by off-resonant excitation of the carrier and first blue sideband [1].

4. Experiment

We work with singly-charged ^{40}Ca$^+$ ions. Due to the presence of the 1.85 T magnetic field, all the energy levels of the ion are split by many GHz. This complicates the experimental requirements considerably as it means that many distinct laser frequencies are required. The energy level diagram is shown in Figure 1. For effective Doppler laser cooling, we require two lasers at 397 nm, four laser frequencies at 866 nm (provided from a single laser using a high-frequency fibre electro-optic modulator), six laser frequencies at 854 nm (provided by a second laser using the same modulator), and one highly-stabilised laser operating at 729 nm for spectroscopy of the narrow qubit transition to the metastable D$_{5/2}$ state. The 854-nm laser is

$4^2P_{3/2}$

+3/2

+1/2

~56GHz/T

−1/2

−3/2

~9.5GHz/T +1/2
−1/2

$4^2P_{1/2}$

393nm $\Gamma_{393}=135\times10^6$

397nm $\Gamma_{397}=132\times10^6$

866nm $\Gamma_{866}=9.1\times10^6$

850nm $\Gamma_{850}=0.96\times10^6$

854nm $\Gamma_{854}=8.5\times10^6$

729nm $\Gamma_{729}=0.9$

$3^2D_{5/2}$

+5/2
+3/2
+1/2
−1/2
−3/2
−5/2

~84GHz/T

$3^2D_{3/2}$

+3/2
+1/2
−1/2
−3/2

~33.5GHz/T

$4^2S_{1/2}$

+1/2

~28GHz/T

−1/2

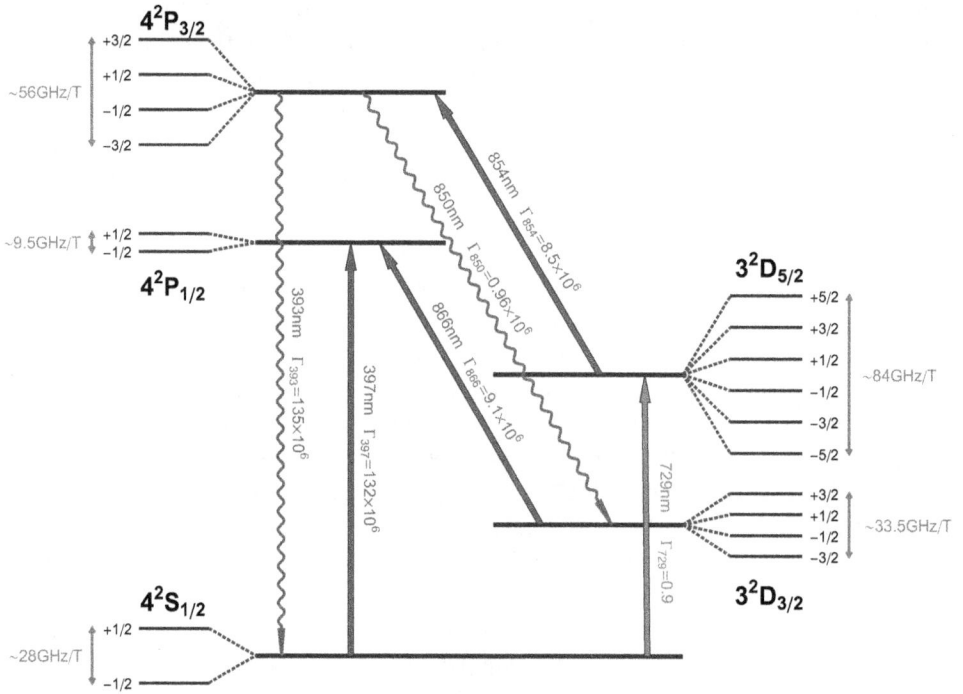

Fig. 1. Energy level diagram for Ca$^+$. The magnetic field splittings are shown greatly exaggerated. Lasers are required at the wavelengths indicated with solid lines.

required for two reasons: firstly because rare decays are possible from the $P_{1/2}$ level to the $D_{5/2}$ level due to magnetic field-induced mixing of the two D levels and also the two P levels [16], and secondly because efficient optical sideband cooling on the 729 nm transition requires the $D_{5/2}$ state to be quenched.

A schematic diagram of the trap is shown in Figure 2. There are laser cooling beams both along the trap axis and in the radial plane. This ensures that all degrees of freedom of the ions are cooled effectively. The radial beam is offset from the centre of the trap as described above. The fluorescence is detected by a photomultipier or imaged onto an EMCCD camera. If two ions are present in the trap, they will both be located on the trap axis when the applied trap potential is low. The residual motion of the ions is much smaller than the size of the laser beam and therefore the size of the images of the ions is limited by the resolution of the optical system. However, different trapping conditions can lead to the ions both lying in the radial plane, in which case they will both rotate around the trap axis because of the presence of the magnetic field. Since they are imaged from the side, this leads to the image of both ions smearing into a line with a length corresponding to the diameter of the orbit. This can be seen in Figure 3 which shows on the left an image of two ions aligned along the vertical trap axis and on the right the same two ions lying in the radial plane.

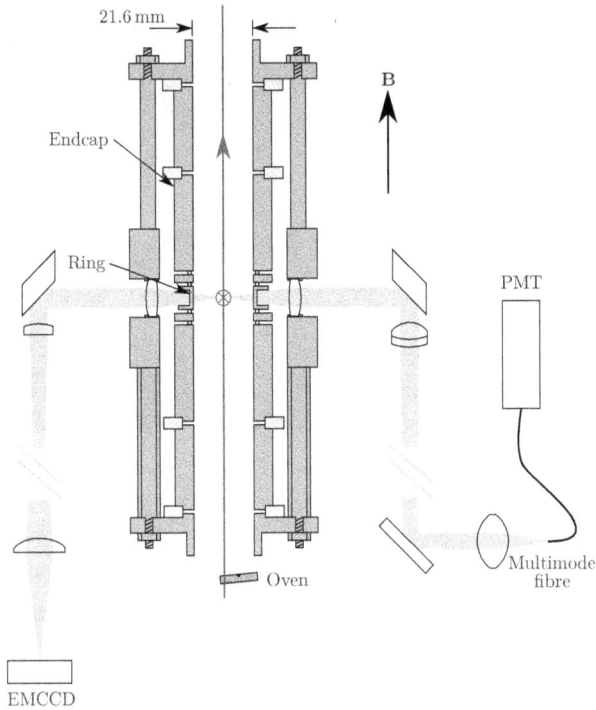

Fig. 2. Schematic diagram of the Penning trap layout showing cooling laser beams and fluorescence detection optics. A cross-section of the cylindrical electrodes is shown. The vertical line is the axial laser cooling beam. The radial cooling beam passes through holes in the ring electrode, points into the page and is marked with a cross. The path taken by the atomic fluorescence is indicated by light shading. The internal diameter of the trap is 21.6 mm. The trap, vacuum enclosure (not shown) and beam-steering optics fit inside the 105-mm vertical bore of the superconducting magnet. EMCCD, electron multiplying charge-coupled device; PMT, photomultiplier tube [10].

5. Optical sideband spectroscopy and cooling of the axial motion

We have recently demonstrated optical sideband spectroscopy of a single ion in the Penning trap [17]. For this work we use a narrow-band laser at 729 nm which drives the electric quadrupole transition from the $S_{1/2}$ ground state to the $D_{5/2}$ metastable state (see Figure 1). The ion is pre-cooled using Doppler cooling lasers as described above, for a period of typically 8 ms. After this the cooling lasers are turned off and the ion is probed with the 729 nm laser for typically 40 μs. The cooling lasers are then turned back on and the fluorescence is recorded (typically for the first 8 ms). If the ion was shelved in the $D_{5/2}$ state, then there is no fluorescence initially. At each laser frequency we repeat this measurement cycle 100 to 400 times in order to obtain the probability of excitation to the metastable level. A plot of this probability as

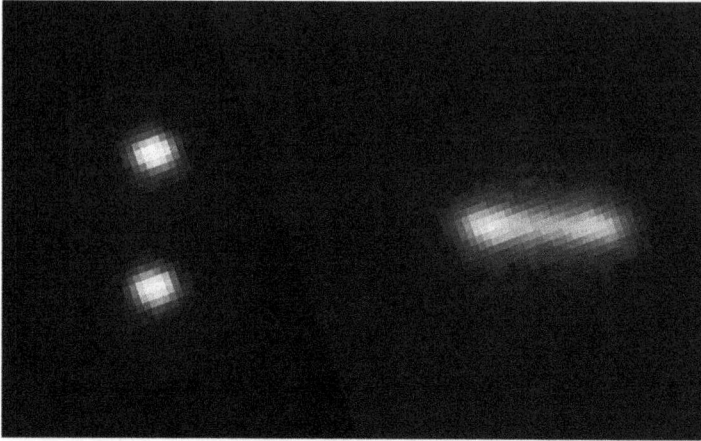

Fig. 3. Two images of two ions aligned along (left) or perpendicular (right) to the vertical trap axis. The separation of the two ions on the left is approximately 30 μm. When the ions are aligned perpendicular to the trap axis, their rotation about the trap axis leads to the image becoming a line rather than separated points.

a function of probe laser frequency shows motional sidebands in the spectrum as expected.

Figure 4 shows a typical spectrum for the axial motion obtained in this way. The measured temperature of 0.47 mK is close to the calculated Doppler limit for this motion. The relatively low value for the axial frequency of the ion (here 400 kHz) leads to a relatively large value for the Lamb-Dicke parameter (0.2) compared to many experiments in RF traps. As a result of this, we observe more than one sideband on either side of the carrier peak. It also means that the initial quantum number for the axial motion prior to sideband cooling (i.e. after Doppler laser cooling) is relatively high (~ 24).

In order to cool the ion beyond the Doppler limit, it is necessary to use optical sideband cooling. For this the 729 nm laser is tuned to the first red sideband of the transition and each time the ion absorbs a photon the quantum number of the axial motion reduces by one. The rate of this process is limited by the spontaneous decay lifetime of the metastable level, which is approximately 1 s, so it is speeded up by weakly irradiating the ion with a laser at 854 nm, which excites the ion from the metastable $D_{5/2}$ level to the $P_{3/2}$ level, from which it can decay rapidly back to the ground state.

In general a sideband cooling period of 20 ms is sufficient to cool to the ground state, but due to the high initial value of $\bar{n}$, there is a small but significant probability that the ion is in a state with $n > 150$. Around this value of n, the amplitude of the first red sideband becomes zero, so cooling on this sideband will not be effective if the initial state is higher than this. In order to avoid population being trapped in this way, it is necessary to employ a sequence of pulses using both the first and second red sidebands for cooling.

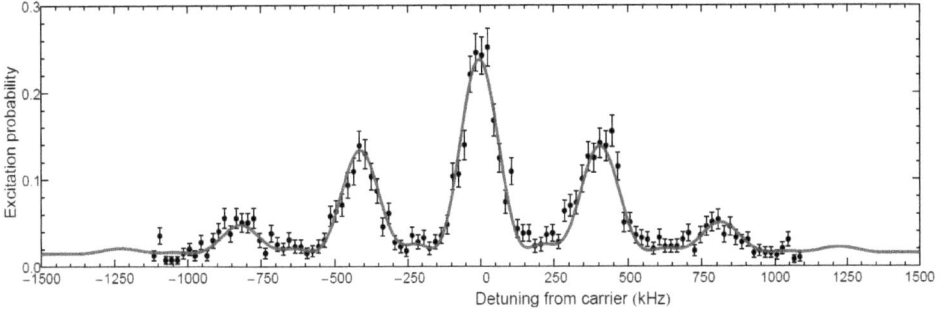

Fig. 4. Optical sideband spectrum of the axial motion of a single ion in a Penning trap after Doppler cooling, recorded at a trap potential of 240 V. The solid line is a fit to the full quantum dynamics of the system and corresponds to a temperature of $T = 0.47 \pm 0.04$ mK ($\bar{n} = 24 \pm 2$). The error bars reflect the statistical uncertainty in the shelving probability arising from the number of repetitions of the measurement cycle at each frequency step [17].

We have now experimentally achieved optical sideband cooling to the ground state of the axial motion of the ion for the first time [18]. A typical spectrum after 20 ms of sideband cooling is shown in Figure 5. The ratio of the amplitude of the first red sideband to the amplitude of the first blue sideband is approximately equal to $\bar{n}$, the average excitation number of the motion. Our final average phonon number of $\bar{n} \sim 0.02$ is consistent with the theoretical estimate of this quantity for our system.

By inserting a waiting time into the pulse sequence between the sideband cooling phase and the spectroscopy phase, we are able to measure the heating rate of the axial motion. We measure the change in $\bar{n}$ with waiting periods of 0, 50 and 100 ms and we find a typical heating rate of $\dot{\bar{n}} = 0.3(2)$ s^{-1} [18].

The measured heating rate in these experiments is very low compared to most measurements in RF traps at room temperature [19]. This is expected because the trap is large compared to typical RF traps, and therefore heating mechanisms related to patch potentials on the electrode surfaces are much less important.

6. Spectroscopy of the radial motion

We have also been able to measure spectra of the radial motion of the ion, which show resolved sidebands due to the cyclotron motion with many magnetron sidebands around each cyclotron peak [17]. From these spectra we are able to make the first direct measurement of the temperature of the radial motion of a laser-cooled ion in a Penning trap. The temperature of the cyclotron motion is typically a few times the expected value, which is of the same order of magnitude as the standard

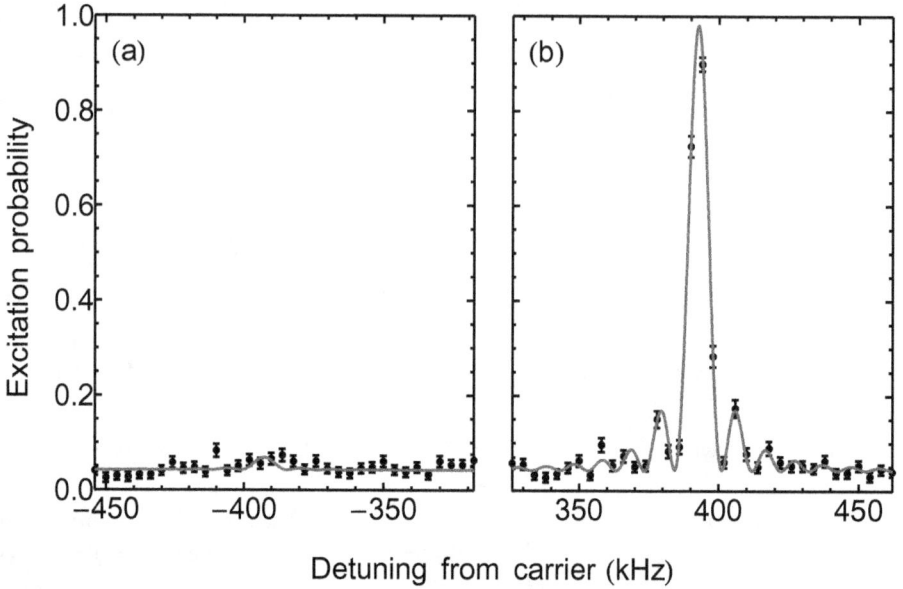

Fig. 5. (a) First red and (b) First blue sidebands after optical sideband cooling, with a trap frequency of $\omega/2\pi = 389$ kHz and a 729-nm probe time of 100 μs. The solid line is a fit to the Rabi dynamics with a constant background, which gives an average phonon number of $\bar{n} = 0.02(1)$ [18].

Doppler limit. However, these spectra demonstrate clearly that the temperature of the magnetron motion is much lower than that of the cyclotron motion. If this were not the case, the magnetron sidebands around each cyclotron sideband would overlap. We find that the magnetron temperature is of the order of tens of microkelvin, much lower than the standard Doppler limit. This is another consequence of the unusual nature of the magnetron motion and in particular the low rate of cooling of this motion, even with the offset radial cooling beam [11] (see Section 3).

We have carried out simulations of the process of sideband cooling of the radial motion. This is more complicated than for the axial motion as there are two degrees of freedom to be cooled at the same time. It turns out that the most efficient way to do this is to cool the two motions in turn rather than to try to cool both at the same time. Note that because the total magnetron energy is negative, it is necessary to sideband cool on the *blue* sidebands of this motion rather than the red sidebands. Also because the initial quantum number of the magnetron motion is typically very high (> 1000), it is more efficient to cool initially on a high-order magnetron sideband rather than on the first sideband so that each absorption removes many phonons rather than just one. We have carried out initial experimental studies of cooling of the radial motion, based on the results of simulations, with encouraging results.

7. Formation of ion Coulomb crystals (ICC)

If the temperature of the trapped ions is reduced far enough, it is possible for an ion Coulomb crystal to form [20]. The so-called Coulomb coupling parameter, $\Gamma = e^2/4\pi\epsilon_0 a_{ws}kT$, is defined to be the ratio of the Coulomb interaction energy between nearest neighbours to the thermal energy. Crystallisation takes place when this parameter becomes greater than about 178 for a large number of particles. Here a_{ws} is the Wigner-Seitz radius, defined by $4\pi a_{ws}^3 n/3 = 1$, with n being the number density of the ions. Such crystals were first reported in a Penning trap by the NIST group in 1988 [21]. Later experiments revealed the structure of the crystals by Bragg scattering of the laser light [22].

Unlike in a radiofrequency trap, the ion crystal in a Penning trap always rotates. This is a consequence of the Lorentz force arising from the presence of the magnetic field B. However, the rotation frequency, ω_R, is not fixed, taking a range of values between ω_m and ω_c'. As is well known, the rotation frequency is linked to the number density of the plasma or crystal through [9]

$$n = 2\epsilon_0 m \omega_R (\omega_c - \omega_R)/e^2. \tag{2}$$

In our case the rate of rotation of the crystal is determined by the parameters of the radial laser cooling beam (specifically, the gradient of the laser intensity at the position of the ion and the laser frequency) [23] and these parameters are not known accurately. The rotation frequency is therefore determined in practice by comparing the experimental images of ICC with simulations that take into account their rotation.

Using our apparatus we are able to trap and image up to approximately 30 ions in a chain along the axis of the trap when we use a low trapping voltage (see [10]). At higher trapping voltages, the ions form into three-dimensional ICC and we are able to manipulate the conformation of the ICC by varying the trapping voltage. At the highest voltages the crystal forms a two-dimensional sheet, as employed by the NIST group in their experiments [6]. A representative series of images of a crystal containing 15 ions is shown in Figure 6. Each of the images obtained experimentally on the left of each pane is compared to a computer simulation on the right of each pane. The simulation program finds the lowest energy configuration of the ions in a three-dimensional ellipsoidal potential well by starting from an initial random arrangement of ions and iteratively moving each ion in the direction of the total force acting on it from all the other ions and the trapping fields, until equilibrium is reached. A simulated image is formed by performing an angular average and taking into account the finite imaging resolution of the optical system. This can then be compared visually with the experimental image. In the simulation, the axial potential is known from the applied trapping voltage. The effective radial potential is a function of the rotation frequency of the crystal. A comparison of each image with simulations obtained for different effective radial potentials therefore allows us to determine the rotation speed of the crystal in each of the images.

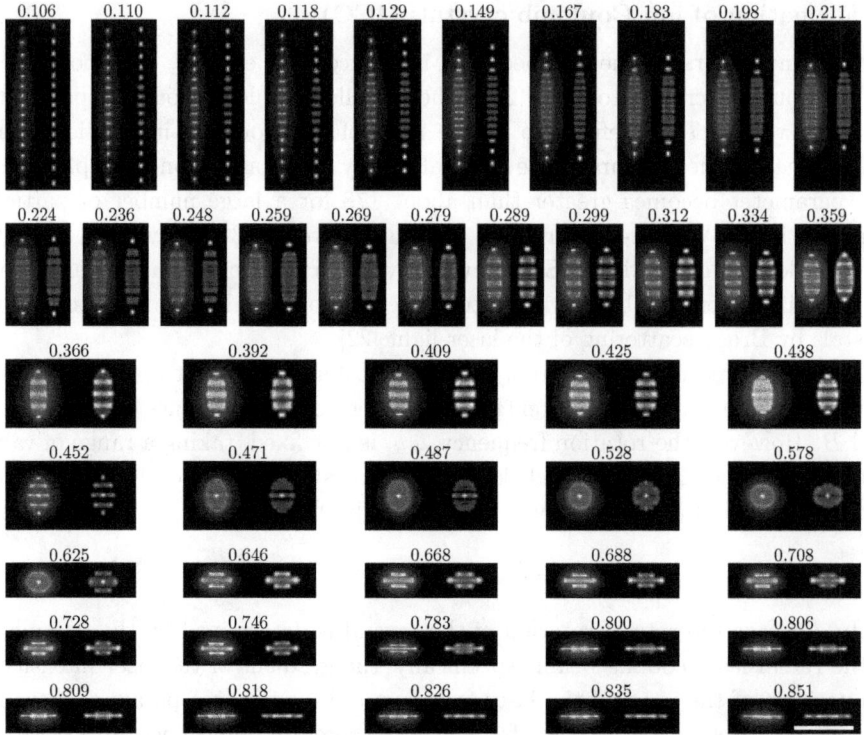

Fig. 6. Experimentally obtained images of 15-ion crystals for different axial potentials (left of each panel), compared with computer simulations (right of each panel). By increasing the axial confinement, a linear string is transformed into a zigzag structure, then a 3D crystal and finally a planar structure. Each image is labelled with the value of the normalized axial trapping frequency (the trap becomes unstable when this quantity is equal to unity). There is a 100-μm scale bar in the bottom right-hand pane, which applies to all the images [10].

The comparison of the experimental images with the computer simulations shows that all the observed crystal conformations are reproduced well in the simulations. Moreover, the crystal rotation speed which is determined in this way from the simulations is roughly constant for different crystal conformations and trapping voltages. In Figure 6 this value is approximately $\omega_c/4$, which is consistent with an estimate based on the known radial cooling laser beam parameters [10].

Small ICC have applications in the areas of quantum information processing and quantum simulation, for example in the simulation of spin frustration in a three-ion system [24]. The Penning trap may offer advantages for this type of work, due to the fact that it allows ICC to be prepared in a variety of one, two, and three-dimensional conformations with known symmetries. As an example of this, an ICC consisting of 6 ions (one in the centre of the trap and five in a ring around it) has been proposed for the demonstration of quantum error-correcting protocols (the five-qubit code and the five-qubit repetition code) using only global operations [25]. The Penning trap is the natural environment in which to perform this experiment.

8. Conclusions

We have demonstrated optical sideband spectroscopy of the radial and axial motions of an ion in a Penning trap. This allows direct measurements to be made of the temperatures of the different degrees of freedom after Doppler cooling. We have shown that the magnetron motion has an equilibrium temperature that is at least an order of magnitude below the standard Doppler limit. We have also demonstrated optical sideband cooling of a single ion in a Penning trap for the first time, achieving a ground state population of approximately 98%. The heating rate was measured to be extremely low compared to other ion trap systems. We have also made a study of the properties of ion Coulomb crystals in a Penning trap. We are able to control the conformation of the crystal over a wide range by varying the trapping potential. The rotation frequency of the crystal can be estimated by matching the image of the crystal to simulations that take into account the effect of the rotation on the radial trapping potential. The combination of controllable conformations of small ICC and the very low heating rate found in our measurements points to the potential for the use of sideband-cooled ions in a Penning trap for applications in quantum information processing or quantum thermodynamics.

Acknowledgments

This work was supported by the UK Engineering and Physical Sciences Research Council (EP/D068509/1) and by the European Commission STREP PICC (FP7 2007-2013 Grant number 249958). We gratefully acknowledge financial support towards networking activities from COST Action MP 1001: Ion Traps for Tomorrows Applications. We also gratefully acknowledge J. Hwang for the loan of the electron multiplying charge-coupled device camera.

References

[1] F. Diedrich, J. C. Bergquist, W. M. Itano and D. J. Wineland, Laser cooling to the zero-point energy of motion, *Phys. Rev. Lett.* **62**, 403 (1989).
[2] D. J. Wineland, R. E. Drullinger and F. L. Walls, Radiation-pressure cooling of bound resonant absorbers, *Phys. Rev. Lett.* **40**, 1639 (1978).
[3] J. J. Bollinger, J. S. Wells, D. J. Wineland and W. M. Itano, Hyperfine structure of the 2p ^{2}P$_{\frac{1}{2}}$ state in ^{9}Be$^+$, *Phys. Rev. A* **31**, 2711 (1985).
[4] R. Thompson, G. Barwood and P. Gill, Laser cooling of magnesium ions confined in a Penning trap, *Optica Acta* **33**, 535 (1986).
[5] Z. Andelkovic, R. Cazan, W. Noertershaeuser, S. Bharadia, D. Segal, R. Thompson, R. Joehren, J. Vollbrecht, V. Hannen and M. Vogel, Laser cooling of externally produced Mg ions in a Penning trap for sympathetic cooling of highly charged ions, *Phys. Rev. A* **87**, 033423 (2013).
[6] B. C. Sawyer, J. W. Britton, A. C. Keith, C. C. J. Wang, J. K. Freericks, H. Uys, M. J. Biercuk and J. J. Bollinger, Spectroscopy and thermometry

of drumhead modes in a mesoscopic trapped-ion crystal using entanglement, *Physical Review Letters* **108**, p. 213003 (2012).

[7] G. Z. K. Horvath, R. C. Thompson and P. L. Knight, Fundamental physics with trapped ions, *Contemporary Physics* **38**, 25 (1997).

[8] K. Blaum, Y. N. Novikov and G. Werth, Penning traps as a versatile tool for precise experiments in fundamental physics, *Contemporary Physics* **51**, 149 (2010).

[9] J. J. Bollinger, J. N. Tan, W. M. Itano, D. J. Wineland and D. H. E. Dubin, Nonneutral ion plasmas and crystals in Penning traps, *Physica Scripta* **T59**, 352 (1995).

[10] S. Mavadia, J. F. Goodwin, G. Stutter, S. Bharadia, D. R. Crick, D. M. Segal and R. C. Thompson, Control of the conformations of ion Coulomb crystals in a Penning trap, *Nature Communications* **4**, p. 2571 (2013).

[11] W. M. Itano and D. J. Wineland, Laser cooling of ions stored in harmonic and Penning traps, *Phys. Rev. A* **25**, 35 (1982).

[12] R. C. Thompson and J. Papadimitriou, Simple model for the laser cooling of an ion in a Penning trap, *Journal of Physics B – Atomic Molecular and Optical Physics* **33**, 3393 (2000).

[13] R. J. Hendricks, E. S. Phillips, D. M. Segal and R. C. Thompson, Laser cooling in the Penning trap: An analytical model for cooling rates in the presence of an axializing field, *Journal of Physics B-Atomic Molecular and Optical Physics* **41**, 035301 (2008).

[14] X. P. Huang, J. J. Bollinger, T. B. Mitchell and W. M. Itano, Phase-locked rotation of crystallized non-neutral plasmas by rotating electric fields, *Physical Review Letters* **80**, 73 (1998).

[15] S. Bharadia, M. Vogel, D. M. Segal and R. C. Thompson, Dynamics of laser-cooled Ca^+ ions in a Penning trap with a rotating wall, *Applied Physics B – Lasers and Optics* **107**, 1105 (2012).

[16] D. R. Crick, S. Donnellan, D. M. Segal and R. C. Thompson, Magnetically induced electron shelving in a trapped Ca^+ ion, *Phys. Rev. A* **81**, p. 052503 (2010).

[17] S. Mavadia, G. Stutter, J. F. Goodwin, D. R. Crick, R. C. Thompson and D. M. Segal, Optical sideband spectroscopy of a single ion in a Penning trap, *Phys. Rev. A* **89**, p. 032502 (2014).

[18] J. F. Goodwin, G. Stutter, R. C. Thompson and D. M. Segal, Sideband cooling an ion to the quantum ground state in a Penning trap with very low heating rate, *Phys. Rev. Lett.* **116**, 143002 (2016).

[19] http://www.quantum.gatech.edu/heating_rate_plot.shtml.

[20] R. C. Thompson, Ion Coulomb crystals, *Contemporary Physics* **56**, 63 (2015).

[21] S. L. Gilbert, J. J. Bollinger and D. J. Wineland, Shell-structure phase of magnetically confined strongly coupled plasmas, *Phys. Rev. Lett.* **60**, 2022 (1988).

[22] J. N. Tan, J. J. Bollinger, B. Jelenkovic and D. J. Wineland, Long-range order in laser-cooled, atomic-ion Wigner crystals observed by Bragg scattering, *Physical Review Letters* **75**, 4198 (1995).

[23] M. Asprusten, S. Worthington and R. C. Thompson, Theory and simulation of ion Coulomb crystal formation in a Penning trap, *Applied Physics B – Lasers and Optics* **114**, 157 (2014).

[24] K. Kim, M. S. Chang, S. Korenblit, R. Islam, E. E. Edwards, J. K. Freericks, G. D. Lin, L. M. Duan and C. Monroe, Quantum simulation of frustrated Ising spins with trapped ions, *Nature* **465**, 590 (2010).

[25] J. F. Goodwin, B. J. Brown, G. Stutter, H. Dale, R. C. Thompson and T. Rudolph, Trapped-ion quantum error-correcting protocols using only global operations, *Phys. Rev. A* **92**, p. 032314 (2015).

Bose-Einstein Condensation of Photons versus Lasing and Hanbury Brown-Twiss Measurements with a Condensate of Light

Julian Schmitt, Tobias Damm, David Dung, Frank Vewinger, Jan Klaers*, and Martin Weitz[†]

Institut für Angewandte Physik, Universität Bonn,
Wegelerstr. 8, 53115 Bonn, Germany
[†] E-mail: martin.weitz@uni-bonn.de
www.iap.uni-bonn.de/ag_weitz/

The advent of controlled experimental accessibility of Bose-Einstein condensates, as realized with e.g. cold atomic gases, exciton-polaritons, and more recently photons in a dye-filled optical microcavity, has paved the way for new studies and tests of a plethora of fundamental concepts in quantum physics. We here describe recent experiments studying a transition between laser-like dynamics and Bose-Einstein condensation of photons in the dye microcavity system. Further, measurements of the second-order coherence of the photon condensate are presented. In the condensed state we observe photon number fluctuations of order of the total particle number, as understood from effective particle exchange with the photo-excitable dye molecules. The observed intensity fluctuation properties give evidence for Bose-Einstein condensation occurring in the grand-canonical statistical ensemble regime.

Keywords: Bose-Einstein condensation, photon gases, condensation dynamics, photon statistics.

1. Introduction

For material particles of integer spin (bosons), Bose-Einstein condensation to a macroscopically occupied ground state minimizes the free energy when cooled to very low temperatures at sufficient density [1]. Other than for gases of material particles [2–6], Bose-Einstein condensation usually does not occur for photons [7]. In the most well-known photon gas, blackbody radiation, photons disappear in the system walls when cooled to low temperature instead of exhibiting Bose-Einstein condensation to the ground mode. This is also expressed by the common statement that the chemical potential for photons vanishes. Early theoretical work has proposed Bose-Einstein condensation of photons in the Compton scattering of X-rays [8], and more recently Chiao proposed a two-dimensional photon fluid in a nonlinear resonator [9]. Besides in blackbody radiation, thermalization effects have long been accounted for in the description of multimode intracavity spectroscopy laser setups below the laser threshold [10]. More recently, Bose-Einstein condensation of exciton-polaritons, which are mixed states of matter and light in the strongly

---

*Present address: Institute for Quantum Electronics, ETH Zurich, Auguste-Piccard-Hof 1, 8093 Zurich, Switzerland

bound limit, has been experimentally achieved [3–5]. Here the material part of the polaritons drive the system into or near thermal equilibrium. Our group in 2010 observed Bose-Einstein condensation of photons in a dye-filled optical microcavity [11], see also recent work by Marelic and Nyman [12]. A very short optical cavity here imprints an effective low-frequency cutoff for photons, with a spectrum of allowed photon energies well above the thermal energy in frequency units. Thermalization of the photon gas to the (rovibrational) temperature of the gas proceeds by absorption and re-emission processes on the dye molecules. For corresponding theoretical works, see [13–20].

This article reviews recent experiments of our group studying the transition from laser-like dynamics to Bose-Einstein condensation of photons upon variation of the thermalization rates of the photon gas to the dye medium [21]. Moreover, we describe work observing grand-canonical number statistics of the condensate emission, as understood from effective particle exchange with the reservoir of photo-excitable dye molecules [18, 19, 22].

In the following, Chapter 2 describes the experimental dye microcavity system, in which the Bose-Einstein condensate is generated, and Chapter 3 presents experiments studying a crossover between laser-like nonequilibrium dynamics and Bose-Einstein condensation of photons. Further, Chapter 4 gives results obtained from studying the condensate number statistics and Chapter 5 concludes this article.

2. Two-dimensional photon gas in microcavity

Our experimental approach utilizes a thermal coupling of a two-dimensional photon gas to a bath of dye molecules. A simplified schematic of the experiment, which has been previously discussed in detail e.g. in Ref. [23], is shown in Fig. 1(a). The experiments are conducted in a cavity consisting of two curved mirrors spaced in the micrometer regime, filled with dye in liquid solution. The mirrors, due to their small spacing, impose an upper limit to the optical wavelength that fits into the cavity, corresponding to a restriction of energies to a minimum cutoff of $\hbar\omega_c \approx 2.1$ eV, which is much larger than thermal energy $k_B T \approx 1/40$ eV at room temperature ($T = 300$ K). In this case, thermal emission of photons into the modes of the cavity is suppressed by a factor of order $\exp(-\hbar\omega_c/k_B T) \approx 10^{-36}$ with the above given numbers, which is a precondition for an independent tuning of photon number and temperature. Note that the limit $\hbar\omega \gg k_B T$ is as well fulfilled in usual laser physics, while this separation of energy scales is not fulfilled for a blackbody radiator. By repeated absorption re-emission processes, the photons thermalize to the rovibrational temperature of the dye, which corresponds to room temperature. In the course of thermalization, the longitudinal modal quantum number of the

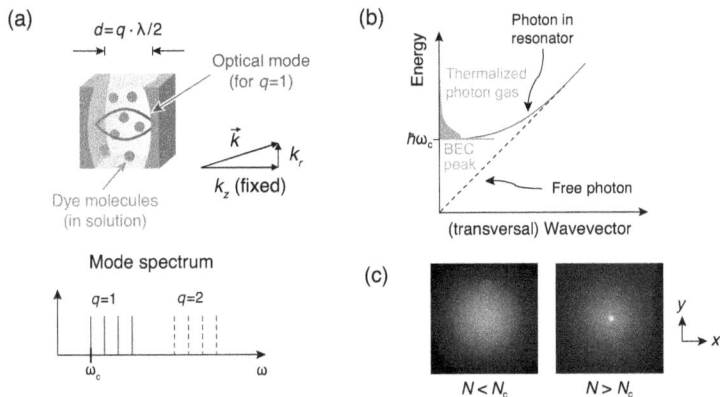

Fig. 1. (a) Scheme of optical microresonator (top) and mode spectrum (bottom) for the case of a spacing between resonator mirrors of half an optical wavelength ($q = 1$). The resonator is filled with dye solution and the photons are thermally coupled to the dye by repeated absorption-emission processes. (b) Photon dispersion in resonator (solid line) and the dispersion of a free photon (dashed line). (c) Spatial images of the cavity emission both in the thermal (left) and in the condensed regime (right), the latter with the bright spot in the center corresponding to the BEC peak.

photons in the thin cavity remains fixed, and the photon dynamics is restricted to the remaining two modal degrees of freedom. In thermal equilibrium, the photon frequencies will then be distributed by $\simeq k_{B}T/\hbar$ above the low-frequency cutoff. Rapid decoherence from collisions with solvent molecules prevents a coupling of the phases of the dye molecular dipole and the photon, so that we can well assume that in our experiments photons instead of polaritons are studied.

Inside the resonator the optical dispersion becomes quadratic, see Fig. 1(b), and the photon gas behaves equivalent to a two-dimensional gas of massive bosons that is harmonically confined, with the latter being caused by the curvature of the cavity mirrors. In contrast to a homogeneous two-dimensional Bose gas, Bose-Einstein condensation here is possible [24]. Interestingly, the effective mass $m_{ph} = \hbar\omega_c/c^2$, where c denotes the speed of light in the medium and ω_c the cutoff frequency, is some 10 orders of magnitude smaller than the mass of alkali atoms, and the Bose-Einstein condensation transition temperature can be at room temperature. In our experiment, an initial photon population is injected into the dye cavity system by pumping the dye with an external laser beam, either in a temporally pulsed or quasi-cw way. In the latter case, losses from the 'photon box' can be compensated for by maintaining the molecular excitation level of the dye at a constant level. We have experimentally observed both the thermalization [25], as well as Bose-Einstein condensation of photons [11]. Figure 1(c) shows typical images for the emitted radiation transmitted through a cavity mirror below (left) and above the phase transition to a Bose-Einstein condensate (right). We observe photon gases

with typically up to 70% condensate fraction very closely following expectations for
a thermal equilibrium distribution. Evidence for a BEC of photons was obtained
from the observed spectra showing Bose-Einstein distributed photon energies with
a macroscopically occupied peak on top of a thermal cloud, the observed threshold
of the phase transition showing the predicted absolute value and scaling with e.g.
mirror curvature, and condensation in the trap center even for a spatially offset
pump beam, as possible by the thermalization [11, 21].

3. Nonequilibrium lasing versus equilibrium condensation of photons

Trapped dilute cold atomic gas systems achieve a state that is very close to that
described by a thermal equilibrium distribution [1, 2]. The establishment of thermal
equilibrium conditions demands that there are separate timescales for thermalization with respect to that of losses and pump terms, being a precondition for the
concept of Bose-Einstein condensation. In the field of exciton-polaritons, where life-
times of the quasiparticles typically are as short as several picoseconds, the question
whether a system that is pumped and exhibits losses can show Bose-Einstein condensation has been discussed [26, 27]. Experiments give evidence for condensation
despite the short lifetime of polaritons [3–5].

    As the here investigated system is situated in the regime of weak coupling of
matter and light, rate equations can be used to describe the dynamics of photons in
the dye microcavity. Usual laser equations, in the limit of negligible loss and assuming that the Kennard-Stepanov law predicting a thermodynamic Boltzmann-type
scaling between the Einstein coefficients for absorption and emission respectively
holds, yield a thermal distribution of photons in the cavity [28]. The intermediate
case of particle equilibrium of the photon gas has in detail been theoretically investigated by Kirton and Keeling [15]. Figure 2 gives a comparison of different states
of cavity photons, assuming $\hbar\omega \gg k_{\mathrm{B}}T$. The left half of the diagram refers to a
pumped medium with no or negligible thermalization, where below laser threshold
the emission is spontaneous, while when inversion is reached at the point when the
gain per cavity round trip exceeds the loss laser operation sets in, yielding a macroscopic occupation of modes independent of energetics [29]. If the cavity lifetime τ_{cav}
exceeds the characteristic thermalization time of the photon gas τ_{th} (right hand side
of diagram), a trapped photon will be reabsorbed in the dye heat bath and thermalize before being lost by e.g. mirror transmission. Below the critical photon
number for Bose-Einstein condensation, again spontaneous processes dominate but
a thermalized distribution of cavity modes is now expected. Once the threshold
for a BEC is reached, the lowest energetic mode will be macroscopically populated,
forming the BEC peak. Both for the laser and the BEC case Bose-enhancement
of modes plays a dominant role in the macroscopic population, a process equally

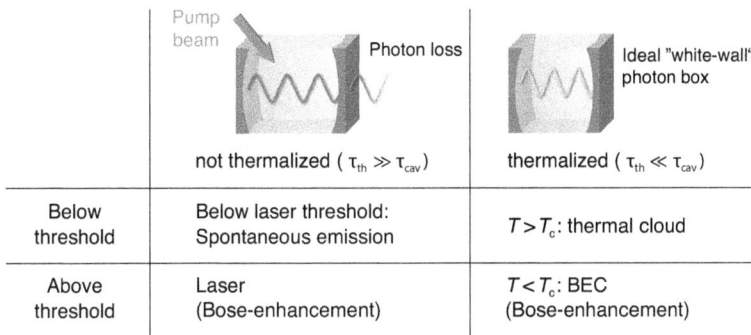

Fig. 2. Classification of optical sources far from and at thermal equilibrium respectively, in the regime of the photon energy $\hbar\omega$, being far above the thermal energy k_BT. We assume that the cavity has a low-frequency cutoff and that the thermalization process conserves the average photon number. Thermal equilibrium is obtained also for a pumped system when a photon thermalizes faster than it is lost by e.g. mirror transmission ($\tau_{th} \ll \tau_{cav}$, right), while in the opposite limit the state remains far from thermal equilibrium, and becomes laser-like when the gain per cavity roundtrip is higher than photon loss ($\tau_{th} \gg \tau_{cav}$, left).

well important in the formation of an atomic condensate, with stimulated scattering processes of atoms in the latter case [30].

We have experimentally determined the characteristic thermalization time in the dye microcavity system by pumping the dye medium with a pulsed picosecond laser and analyzing the subsequent cavity emission in a time-resolved way using a streak-camera [21]. Under spatially homogeneous excitation of the dye molecules the temporal evolution of the spectral photon distribution reveals the thermalization time after which the spectrum is described by a 300K Bose-Einstein distribution, which is of order of the reabsorption timescale (for typical parameters approximately 20 ps) of a photon in the dye medium. This indicates that indeed thermalization occurs to the molecular bath. Further, we studied the transition between laser dynamics and equilibrium condensation of photons, for which a focused pump pulse was irradiated spatially removed from the trap center, as to energetically remove the initial state of cavity photons from the low frequency ground state, see Fig. 3 for corresponding data. In the course of our experiments, the cavity cutoff was varied in order to tune the coupling strength to the dye bath, allowing to control the thermalization time of the photons. The data shown on the left hand side of Fig. 3 was recorded for a position of the cavity cutoff $\lambda_c = 596$ nm, for which the wavelength of cavity photons is detuned relatively far from the dye zero-phonon line, so that dye absorption and thermal contact of photons to the dye is weak. We experimentally observe the oscillation of cavity photons within the trapping potential imprinted by the mirror curvature. The photons remain in a non-thermal state at high energies far above the cutoff-frequency, and the according dynamics resembles an analogue to (mode-locked) laser oscillation. Photons here leak out of

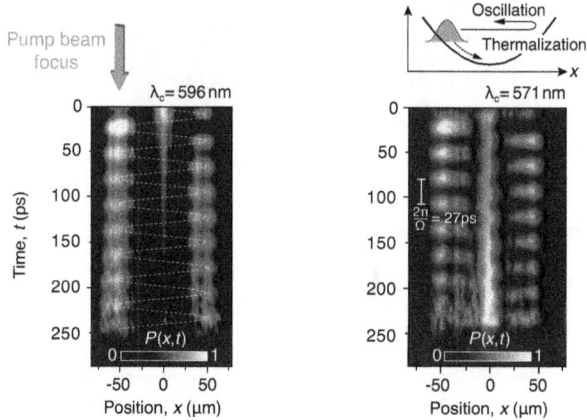

Fig. 3. Temporal evolution of the spatial profile (line normalized) of the radiation transmitted through one cavity mirror for two cutoff wavelengths. A pump beam with 27 μm diameter spatially displaced by 50 μm from the trap center excites an optical wave packet oscillating in the harmonic trapping potential (indicated on top, right). For the data recorded with shorter cutoff wavelength (right) and correspondingly larger dye absorption a photon condensate in the trap center gradually builds up, while for low reabsorption no condensate emerges (left).

the cavity before they have a chance to thermalize. The right hand side of Fig. 3 shows data recorded for a cavity cutoff $\lambda_c = 571$ nm, tuning cavity photons in a wavelength range where they are reabsorbed from the traversing wave packet by dye molecules and thermal contact is established. We observe that with advancing times the photon gas accumulates in the trap center and here forms a Bose-Einstein condensate. The observations are consistent with further measurements of the time evolution of the photon spectra. While for weak coupling to the dye heat bath the photon spectrum persists in its initial nonequilibrium state, a spectral redistribution of fluorescence to an equilibrium Bose-Einstein distribution is observed, when thermal contact between photons and dye molecules is enhanced [21].

4. Hanbury brown-twiss measurements of the photon condensate

The above described studies have demonstrated radiative coupling of photons to a thermal heat bath in the composite dye-photon system. Evidently, the photon gas can not be considered as a system isolated from its environment in the sense of the microcanonical statistical ensemble. This leads us to the question, whether the electronically excited dye molecules can additionally constitute an effective reservoir species for the photons, allowing for particle exchange among the two subsystems. In the present section, we describe corresponding experimental work determining the intensity correlations of the dye microcavity emission. The main result is that we observe number fluctuations in the condensed state which are of same order as the average particle number of the condensate. This gives evidence for the condensation occurring in a system described by grand-canonical statistical conditions, due to possible effective particle exchange with between photons and dye electronic excitations.

In general, different statistical ensembles represent different conservation laws that can be realized in nature. The microcanonical and canonical ensemble, respectively, refer to physical systems with a fixed number of particles, while energy is fixed in the former case and allowed to fluctuate around a mean value, determined by contact to a heat reservoir, in the latter case. In the grand-canonical ensemble both energy and particle number of a system can vary due to contact with a particle and energy reservoir and one here finds relative fluctuations of all single particle levels of 100%. For most problems in statistical physics one assumes that the different statistical ensembles become interchangeable in the thermodynamic limit, meaning that relative fluctuations vanish, i.e. $\Delta N / \bar{N} \to 0$, with $\bar{N}$ as the average total particle number and ΔN its rms fluctuations. Notably, the ideal Bose gas represents an exemption from this generalization, as assuming grand-canonical conditions for the case of the macroscopically occupied ground state present in the Bose-Einstein condensed case yields statistical fluctuations of order of the total particle number, i.e. $\Delta N \approx \bar{N}$. The fluctuations here do not freeze out at low temperature, instead the prediction is just the opposite: the size of the fluctuations approaches the average particle number as the condensate fraction reaches unity. This counterintuitive phenomenon is commonly referred to as the 'grand-canonical fluctuation catastrophe' [31–33].

Grand-canonical ensemble conditions do not apply to a cloud of an ultracold atomic gases well isolated from the environment, as well as for present polariton condensation experiments [3–5]. The situation however is less obvious for a spatially finite region within a large reservoir, as was first discussed in relation to e.g. liquid helium systems. The physical significance of the grand-canonical ensemble in the condensed phase has long been an open question. Ziff, Uhlenbeck and Kac showed that, for a system in diffusive contact with a spatially separated particle reservoir, the grand-canonical ensemble loses its validity [32]. These arguments however do not necessarily apply for other types of reservoirs. In the here discussed dye microcavity system, the dye molecules act both as a heat bath and a particle reservoir for the photon gas in a grand-canonical sense, with the contact between system and reservoir not being realized diffusively but by absorption and emission processes, in a spatially overlapping geometry. We have predicted grand-canonical number fluctuations for this system [18], and the obtained theory results for the photon number distribution have been confirmed [19].

To experimentally observe the intensity correlations of the condensate mode of dye microcavity emission, we use a Hanbury Brown-Twiss setup [22]. This experi-

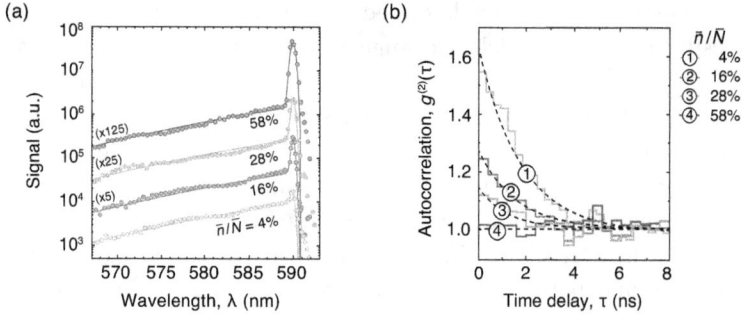

Fig. 4. (a) Spectral photon distribution for different condensate fractions (circles), each following a 300 K Bose-Einstein distribution (solid lines). Curves have been vertically shifted for clarity. (b) Corresponding second-order correlation functions $g^{(2)}(\tau)$ show photon bunching up to high condensate fractions. Experimental parameters: condensate wavelength $\lambda_c = 2\pi c/\omega_c = 590$ nm ($\hbar\Delta = -6.7\,k_B T$), dye concentration $\rho = 10^{-3}$ mol/l (rhodamine 6G).

ment is carried out by pumping dye microcavity with typically 150 ns long pulses derived by acousto-optically chopping the emission of a cw laser, which is much longer than the picosecond thermalization time [21]. The experiment thus is operated in a 'quasi'-cw mode; as for the case for our initial works described in Section 2. Part of the radiation transmitted through one cavity mirror is spatially filtered in the far-field Fourier plane to separate the condensate mode from the higher transverse modes. Subsequently, the filtered radiation (condensate mode) is split and directed onto two single-photon avalanche photodiodes. A correlation system records time histograms of detection events at the detectors, from which intensity correlations $g^{(2)}(\tau)$ are determined.

Typical measurement results for a fixed size of the molecular reservoir are presented in Fig. 4. The thermodynamic state of the photon gas is determined by recording spectra as shown in Fig. 4(a) using a fraction of the light emitted from the cavity. The shown spectra all are in the condensed phase, with an average photon number beyond the critical particle number of $N_c = 85\,000$, and show a peak at the wavelength of the cavity cutoff along with a thermal cloud at lower wavelengths. The condensate fraction is obtained by a fit to a 300 K Bose-Einstein distribution. Results for the second-order correlation function $g^{(2)}(\tau)$ are shown in Fig. 4(b). While an immediately second-order coherent correlation signal, with $g^{(2)}(0) = 1$, would be expected above the BEC transition ($N \geq N_c$) in case of a strictly conserved particle number, we observe photon bunching, with $g^{(2)}(0) > 1$, to extend clearly into the condensed phase regime. For large delays, the observed bunching of the condensate light decays. The expected crossover between the fluctuating regime and the regime with second-order coherence is at

$$\bar{n}^2 \simeq \frac{M}{\left(1 + e^{\hbar\Delta/k_B T}\right)\left(1 + e^{-\hbar\Delta/k_B T}\right)} = M_{\text{eff}}, \tag{1}$$

	$\hbar\Delta/k_BT$	ρ(Mol/l)	$M_{\text{eff,Ri}}/M_{\text{eff,R1}}$
R1	-7.8	10^{-4}	1
R2	-7.4	10^{-3}	15
R3	-5.3	10^{-3}	120
R4	-1.8	10^{-4}	300
R5	-2.3	10^{-3}	2000

Autocorrelation, $g^{(2)}(0)$

Fluctuations, $\Delta n/\bar{n}$ (%)

Condensate fraction, $\bar{n}/\bar{N}$ (%)

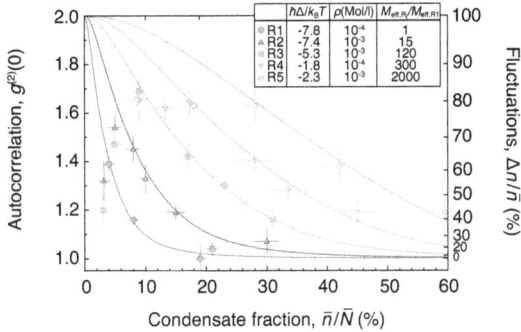

Fig. 5. Zero-delay autocorrelations $g^{(2)}(0)$ versus condensate fraction $\bar{n}/\bar{N}$ for five different reservoirs of relative effective size $M_{\text{eff,R}_i}/M_{\text{eff,R1}}$ with respect to the smallest reservoir R1. Condensate fluctuations extend deep into the condensed phase for high dye concentration ρ and small dye-cavity detuning Δ. Results of a theoretical modeling are shown as solid lines. Experimental parameters: condensate wavelength $\lambda_c = 2\pi c/\omega_c = \{598, 595, 580, 598, 602\}$nm for data sets R1-R5; dye concentration $\rho = \{10^{-4}, 10^{-3}, 10^{-3}\}$mol/l for R1-R3 (rhodamine 6G) and $\rho = \{10^{-4}, 10^{-3}\}$mol/l in R4 and R5 (perylene red).

where M denotes the density of dye molecules, $\bar{n}$ the average number of photons in the condensate mode and $\Delta = \omega_c - \omega_{\text{zpl}}$ the frequency detuning of the condensate from the position of the zero-phonon line of the dye [18, 22].

Figure 5 shows the dependence of the measured zero-delay second-order coherence function $g^{(2)}(0)$ on the condensate fraction for five different combinations of dye concentration and dye-cavity detuning Δ, which varies the effective reservoir size M_{eff} defined in eq. (1). The data sets labelled with R1-R3 have been obtained with rhodamine 6G dye (zero-phonon line at $\omega_{\text{zpl}} \simeq 2\pi c/(545 \text{ nm})$). For measurements R4 and R5, we have used perylene red dye ($\omega_{\text{zpl}} \simeq 2\pi c/(585 \text{ nm})$), which allows us to reduce the detuning between condensate and dye reservoir. For the lowest dye concentration and largest detuning (R1), the particle reservoir is so small that the condensate fluctuations are damped soon above the onset of Bose-Einstein condensation (i.e. when the ground mode population becomes macroscopic). The observed emergence of second-order coherence here is attributed to canonical statistical ensemble conditions present in the system. However, by increasing the dye concentration and decreasing the dye-cavity detuning one can systematically extend the regime of large fluctuations to higher condensate fractions. For the largest reservoir realized (R5), where the effective reservoir is approximately 2000 times larger than for reservoir R1, we observe zero-delay correlations of $g^{(2)}(0) \simeq 1.2$ at a condensate fraction of $\bar{n}/\bar{N} \simeq 0.6$. The condensate here still performs large relative fluctuations, with $\Delta n/\bar{n} = \left(g^{(2)}(0) - 1\right)^{1/2} \simeq 0.45$, although its occupation number is already comparable to the total photon number. We attribute this as clear evidence that the photon statistics in this system is determined by grand-canonical particle exchange between condensate and dye reservoir. The data agree well with predictions based on a theoretical model shown by solid lines, except when the condensate fraction becomes very small (below some 5%). In the latter case the measured correlation function reduces towards smaller values, which is attributed to residual light from thermal cavity modes reaching the detection system because of imperfect mode filtering. The averaging over the large number of modes reduces the observed value for $g^{(2)}(0)$ towards unity.

5. Conclusions

We have described recent experiments with a Bose-Einstein condensate of photons realized in a dye-filled microcavity. The degree of thermalization of the photon gas was varied by tuning the cavity photons closer to resonance or far from resonance with the dye molecules, which in the latter case results in a suppression of thermal contact to the dye and the photons leaking out of the cavity due to mirror losses before they can thermalize, which is reminiscent to usual laser operation. On the other hand, for an enhanced thermal contact with the dye by absorption re-emission processes, photons thermalize to low energetic states near the cavity cutoff and form a Bose-Einstein condensate.

In a further series of experiments, the intensity correlations of the photon condensate generated in the dye microcavity system were determined. Relevant to those measurements is that the photo-excitable dye molecules do not only act as a heat bath, but also as an effective particle reservoir for the photon gas. When the dye reservoir is sufficiently large with respect to the system size, we observe grand-canonical statistical fluctuations in the condensed state, while the fluctuations reduce to the usual Poissonian case for a smaller relative size of the reservoir. The results give evidence for Bose-Einstein condensation in the grand-canonical statistical regime.

An investigation of the first-order coherence properties of the condensate for different statistical ensemble regimes is subject to current experimental studies. For the future it will be important to test for superfluidity of the photon condensate. An intriguing perspective is the investigation of quantum many-body states in photonic lattices, in which cooling alone can allow for the preparation of entangled many-body states in a thermal equilibrium process when the many-body state is the system ground state.

Acknowledgments

We thank M. Fleischhauer for discussions concerning the laser-BEC comparison of Fig. 2, and acknowledge funding from the ERC (INPEC) and the DFG (We 1748-17).

References

[1] See, e.g.: A. J. Leggett, *Quantum liquids* (Oxford University Press, New York, 2006).

[2] See, e.g.: K. Bongs, and K. Sengstock, Physics with coherent matter waves, *Rep. Prog. Phys.* **67**, 907 (2004)

[3] H. Deng, G. Weihs, C. Santori, J. Bloch, and Y. Yamamoto, Condensation of semiconductor microcavity exciton polaritons, *Science* **298**, 199 (2002).

[4] J. Kasprzak, M. Richard, S. Kundermann, A. Baas, P. Jeambrun, J. M. J. Keeling, F. M. Marchetti, M. H. Szymańska, R. André, J. L. Staehli, V. Savona, P. B. Littlewood, B. Deveaud and L. S. Dang, Bose-Einstein condensation of exciton polaritons, *Nature* **443**, 409 (2006).

[5] R. Balili, V. Hartwell, D. Snoke, L. Pfeiffer, and K. West, Bose-Einstein condensation of microcavity polaritons in a trap, *Science* **316**, 1007 (2007).

[6] S. O. Demokritov, V. E. Demidov, O. Dzyapko, G. A. Melkov, A. A. Serga, B. Hillebrands, and A. N. Slavin, Bose-Einstein condensation of quasi-equilibrium magnons at room temperature under pumping, *Nature* **443**, 430 (2006).

[7] See, e.g.: K. Huang, *Statistical Mechanics, 2nd edn* (Wiley, New York, 1987), pp. 293-294

[8] Y. B. Zel'dovich and E. V. Levich, Bose condensation and shock waves in photon spectra, *Sov. Phys. JETP* **28**, 1287 (1969).

[9] R. Y. Chiao, Bogoliubov dispersion relation for a 'photon fluid': Is this a superfluid?, *Opt. Comm.* **179**, 157 (2000).

[10] V. R. Mironenko and V. I. Yudson, Quantum statistics of multimode lasing and noise in intracavity laser spectroscopy, *Sov. Phys. JETP* **52**, 594 (1980).

[11] J. Klaers, J. Schmitt, F. Vewinger, and M. Weitz, Bose-Einstein condensation of photons in an optical microcavity, *Nature* **468**, 545 (2010).

[12] J. Marelic and R. A. Nyman, Experimental evidence for inhomogeneous pumping and energy-dependent effects in photon Bose-Einstein condensation, *Phys. Rev. A* **91**, 033813 (2015).

[13] A.-W. de Leeuw, H. T. C. Stoof, and R. A. Duine, Schwinger-Keldysh theory for Bose-Einstein condensation of photons in a dye-filled optical microcavity, *Phys. Rev. A* **88**, 033829 (2013).

[14] D. W. Snoke and S. M. Girvin, Dynamics of phase coherence onset in bose condensates of photons by incoherent phonon emission, *J. Low Temp. Phys.* **171**, 1 (2013).

[15] P. Kirton and J. Keeling, Nonequilibrium model of photon condensation, *Phys. Rev. Lett.* **111**, 100404 (2013).

[16] A. Kruchkov, Bose-Einstein condensation of light in a cavity, *Phys. Rev. A* **89**, 033862 (2014).

[17] J.-J. Zhang, Y.-H. Yuan, J.-P. Zhang, and Z. Cheng, Temperature dependence of atomic decay rate induced by the BEC of photons, *Physica E* **45**, 177 (2012).

[18] J. Klaers J. Schmitt, T. Damm, F. Vewinger, and M. Weitz, Statistical physics

of Bose-Einstein-condensed light in a dye microcavity, *Phys. Rev. Lett.* **108**, 160403 (2012).

[19] D. N. Sobyanin, Hierarchical maximum entropy principle for generalized superstatistical systems and Bose-Einstein condensation of light, *Phys. Rev. E* **85**, 061120 (2012).

[20] M. C. Strinati and C. Conti, Bose-Einstein condensation of photons with nonlocal nonlinearity in a dye-doped graded-index microcavity, *Phys. Rev. A* **90**, 043853 (2014).

[21] J. Schmitt et al., Thermalization kinetics of light: From laser dynamics to equilibrium condensation of photons, *Phys. Rev. A* **92**, 011602 (2015).

[22] J. Schmitt, T. Damm, D. Dung, F. Vewinger, J. Klaers, and M. Weitz, Observation of grand-canonical number statistics in a photon bose-einstein condensate, *Phys. Rev. Lett.* **112**, 030401 (2014).

[23] J. Klaers, The thermalization, condensation and flickering of photons, *J. Phys. B* **47**, 243001 (2014).

[24] V. Bagnato and D. Kleppner, Bose-Einstein condensation in low-dimensional traps, *Phys. Rev. A* **44**, 7439 (1991).

[25] J. Klaers, F. Vewinger, and M. Weitz, Thermalization of a two-dimensional photonic gas in a "white wall" photon box, *Nature Phys.* **6**, 512 (2010).

[26] M. Wouters, I. Carusotto, and C. Ciuti, Spatial and spectral shape of inhomogeneous nonequilibrium exciton-polariton condensates, *Phys. Rev. B* **77**, 115340 (2008).

[27] H. Deng, D. Press, S. Götzinger, G. S. Solomon, R. Hey, K. H. Ploog, and Y. Yamamoto, Quantum degenerate exciton-polaritons in thermal equilibrium, *Phys. Rev. Lett.* **97**, 146402 (2006).

[28] J. Klaers J. Schmitt, T. Damm, F. Vewinger, and M. Weitz, Bose-Einstein condensation of paraxial light, *Appl. Phys. B* **105**, 17 (2011).

[29] See, e.g.: A. E. Siegman, Lasers (University *Science* Books, 1986)

[30] M. D. Lee and G. W. Gardiner, Quantum kinetic theory. VI. The growth of a Bose-Einstein condensate, *Phys. Rev. A* **62**, 033606 (2000).

[31] I. Fujiwara, D. ter Haar, and H. Wergeland, Fluctuations in the population of the ground state of Bose systems, *J. Stat. Phys.* **2**, 329 (1970).

[32] R. M. Ziff, G. E. Uhlenbeck, M. Kac, The ideal Bose-Einstein gas, revisited, *Phys. Rep.* **32**, 169 (1977).

[33] M. Holthaus, E. Kalinowski, K. Kirsten, Condensate fluctuations in trapped Bose gases: Canonical vs. microcanonical ensemble, *Ann. Phys.* **270**, 198 (1998).

Author Index

www.ingramcontent.com/pod-product-compliance
Lightning Source LLC
Chambersburg PA
CBHW081110220326
41598CB00038B/7297